ACCELERATOR
INSTRUMENTATION

AIP
CONFERENCE
PROCEEDINGS 212

ACCELERATOR
INSTRUMENTATION
UPTON, NY 1989

EDITORS:
EDWARD R. BEADLE
VINCENT J. CASTILLO
BROOKHAVEN NATIONAL
LABORATORY

American Institute of Physics   **New York**

L.C. Catalog Card No. 90-55838
ISBN 0-88318-645-4
DOE CONF 8910165

Printed in the United States of America.

CONTENTS

FOREWORD

Accelerator Conferences by necessity cover a wide range of topics and therefore presentations are limited to the recent advances in the various areas of endeavor. Historical and tutorial treatments, if included, are minimal and that can be frustrating for new practitioners in the field of Accelerator engineering.

The first Accelerator Instrumentation Workshop addresses this pedagogical deficiency along with the continuing need for cooperation among all Accelerator laboratories and Industrial concerns that cater to these laboratories in facing common economic, engineering and technological challenges.

We thank the Speakers (and their supporting staff) for the tremendous effort they put into their contributions at the workshop and the subsequent documentation for these proceedings. We thank the participants for making every session fruitful particularly the square "round-table" conference. It was at this forum that the problems in the field of Accelerator Instrumentation were aired and the first salvo of solutions fired. We thank the secretaries: Joan Depken and Rae Bailey for handling the many details that make workshops possible. We thank the members of the local organizing committees: Arrangements (equipment, meeting-place, housing, refreshments, banquet, and registration); Finance and Publication, all of whom gave unstintingly of their time to make the workshop a resounding success. We thank the Organizing Committee particularly the Co-Chairmen: Dr. G. Bennett and Dr. R. Witkover on whom the burden of the enterprise rested; this forerunner has adequately set the pace for future Accelerator Instrumentation Workshops.

Finally, we express our deepest appreciation for the funding of the workshop to Dr. Derek Lowenstein, who also opened the program, and Dr. David Sutter, Chief of the Advanced Technology R&D Branch, Department of Energy.

We, the Co-Editors of the first annual Accelerator Instrumentation Workshop, feel singularly honored with this assignment.

Co-Editors:
Edward Beadle
Vincent Castillo

ORGANIZING COMMITTEE

G. Bennett	Brookhaven National Laboratory
J. Hinkson	Lawrence Berkeley Laboratory
R. Juras	Oak Ridge National Laboratory
A. Rauchas	Argonne National Laboratory
M. Ross	Stanford Linear Accelerator Center
R. Shafer	Los Alamos National Laboratory
G. Stover	Lawrence Berkeley Laboratory
O. van Dyck	Los Alamos National Laboratory
R. Witkover	Brookhaven National Laboratory

LOCAL ORGANIZING COMMITTEE

E. Beadle
J. Bittner
V. Castillo
S. Hobbie
R. Lankshear
R. McKenzie-Wilson
R. Nawrocky
J. Rothman
A. Stillman
N. Tsoupas
E. Zitvogel
M. Zucker

WORKSHOP CO-CHAIRMEN

G. Bennett
R. Witkover

WORKSHOP SECRETARIES

Rae Bailey
Joan C. Depken, CPS

PROCEEDINGS SECRETARY

Joan C. Depken, CPS

BEAM CURRENT MONITORS

Notes for Workshop on Accelerator Instrumentation,

Brookhaven National Laboratory, October, 1989.

Richard Talman

SSC Laboratory, Dallas, Texas.

ABSTRACT

Instruments for monitoring the beam currents in accelerators are described, with emphasis on non-destructive monitors that do not intercept the beam. The spectral content observed on the output of such devices is analysed for various bunch profiles. Stripline detectors, beam current transformers, and direct current transformers are described and analysed.

1. Introduction

Current or intensity monitors can be separated into destructive and non-destructive varieties, with the latter being the main topic of these notes. To begin with though, a few destructive types will be described.

Monitors can also be differentiated as being primarily for relative or for absolute measurements, with the latter obviously being harder. Some applications for which relative measurements are adequate are determination of the relative size of two or more bunches and determination of the lifetime of a slowly decaying beam in a storage ring. Another common requirement is to determine the efficiency of injection from a transfer line into an accelerator, or inversely, extraction from an accelerator into a transfer line. For these measurements the relative calibration of two monitors, one in the line, one in the accelerator, is what matters. If the two detectors are of identical design it may be possible to measure efficiencies accurately even if neither detector is calibrated absolutely.

Intensity monitors can also be differentiated on the basis of the particular electromagnetic interaction employed. The front end, or pick-up of the monitor can be an electrode, responding to the electric field of the beam; it can be a loop or transformer, responding to the magnetic field; or, at higher frequencies, it can be a resonator or cavity, fully electromagnetic in nature. Of these, the ones to be emphasized here are striplines, which respond to both electric and magnetic fields, but require only an elementary field analysis, and transformers, which respond to magnetic fields. Ultra-high frequency detectors that depend on resonant cavity analysis are described in an accompanying lecture by Bob Webber.

Current monitoring requires nothing more than an ammeter, but naturally the processing circuitry depends on the time scale of interest for the measurement. To measure the longitudinal profile of a short, relativistic bunch of particles, current measurements lasting tens of picoseconds are needed; to measure the slow decay of a circulating beam in a storage ring, current measurements lasting for a day or more may be required. Since these extremes span some sixteen orders of magnitude, it should not be surprising that different techniques are called for.

2. Destructive Intensity Monitors

Most accelerators are repetitive devices that are more or less stable from pulse to pulse. To measure the intensity, occasional pulses can be diverted and dumped in a destructive intensity monitor. If the beam line is for experiments, it passes through physics targets before getting to the monitor. For very weak particle beams various counting techniques are employed to measure the intensity; they do not fall within the scope of these notes. For intense beams a "thick" detector, with output proportional to the intensity, is used. The beam is almost completely absorbed in such a monitor. Often only crude absolute accuracy (say

$\pm$ 10 percent before more accurate empirical calibration) is needed. In that case they can be quite simple and still achieve relative accuracies at the $\pm$ 1 percent level. Some examples will now be given.

Faraday Cup. To measure a current of moving charges, it is natural to convey the charges through the metering wire of an ammeter. If the charges are in vacuum, as is typical in an accelerator, they must first be captured on an anode. (Let us assume that negatively charged electrons are being measured.) If the electrons are energetic, the anode had better be pretty thick to absorb them. For electrons in the GeV range this might require 40 cm of copper. For energies higher than that the method is less practical. As with vacuum tubes, care must be taken that current does not flow to other nearby electrodes. For that reason the absorbing block is biased sufficiently positive (typically in the ten to hundred volt range) relative to nearby surfaces.

Wilson Quantameter: an instrument invented by R. R. Wilson, far more sensitive than a Faraday cup, can be used to measure the total energy content (rather than the current) of either an electron or a photon beam; it is rarely used for proton or other hadron beams. This measures the ionization produced by the electromagnetic showers produced by individual beam particles; this ionization is quite accurately proportional to the total energy, and the proportionality constant is almost the same for electrons and photons. This instrument, roughly cubical, 30 cm on a side, has a multilayered structure, with alternating gaseous-argon-filled gaps and copper plates. Every second plate is grounded and the others are biased to a voltage of some tens of volts, enough to collect ions produced in the argon.

When high energy photon beams were first being used, enormous effort went into calibrating these devices. Experimentally gas purity, pressure, and temperature regulation were important; so was the precision measurement of small quantities of charge. The theoretical calculation of the absolute calibration was essential and accuracies of one or two percent were achieved. This was an instance of "necessity being the mother of invention" as the calculation is not straightforward. It is said that Robert Wilson was the first person to use a Monte

Carlo calculation; he did it to predict electromagnetic shower development. His "computer" was a spinner (as in a child's game) having stopping bands of width proportional to the probabilities of the various atomic processes.

A serious disadvantage of the quantameter is that it saturates at high beam intensity, owing to the recombination of the ions in the argon before they are collected.

Secondary Emission Quantameter. To overcome the saturation problem just mentioned, the secondary emission quantameter was introduced. This has mechanical design much like the Wilson quantameter, except that it is evacuated, not argon-filled. The stopping beam causes secondary electron emission from the surfaces of the plates. These electrons are collected and the charge measured. This monitor is much less sensitive than the original quantameter, and its calibration constant cannot be dead-reckoned, but it responds linearly up to much higher beam intensities.

With care in signal processing, the stopping detectors described so far have been shown to be capable of current sensing up to very high frequencies, approaching the gigahertz range for low energy electron beams.

3. Time Domain and Frequency Domain
Response of Current Monitors

Single Particle. Let s stand for a tangential distance coordinate in a circular accelerator of circumference C_o. A particle of charge e, traveling at speed v_o, on the central orbit will pass a fixed point (call it $s = 0$) at regular intervals of time of length $T_o = C_o/v_o$. The line charge density, per unit length, corresponding to a single passage of the particle, at $t = 0$, is

$$\lambda = e\delta(s) = \frac{e}{|ds/dt|}\delta(t) = \frac{e}{v_o}\delta(t). \tag{3.1}$$

Adding all the passages yields

$$\lambda = \frac{e}{v_o}\sum_{l=-\infty}^{\infty}\delta(t - lT_o). \tag{3.2}$$

This is a "comb" of equally spaced equal strength lines in the time domain. It can also be represented as a sum of terms having sinuisoidal time variation, using the easily derived Fourier series relationship

$$\sum_{l=-\infty}^{\infty}\delta(t - lT_o) = \frac{1}{T_o}\sum_{n=-\infty}^{\infty}\cos 2\pi nt/T_o. \tag{3.3}$$

Defining a "fundamental" oscillation $\cos\omega_o t$ where $\omega_o = 2\pi/T_o$, the current signal can be regarded as the superposition of "harmonics" of the fundamental,

$$\lambda = \frac{e}{v_o T_o}\sum_{n=-\infty}^{\infty}\cos n\omega_o t. \tag{3.4}$$

In these equations λ is that quantity which, when multiplied by a spatial interval ds, yields the charge contained in ds. For example, as a check, calculation of the total (i.e. using the $n=0$, DC term) charge from (3.4) yields $\int_0^{C_o}\lambda ds = e$.

The Fourier series (3.4) can be replaced by an integral over a frequency variable ω, that is as a Fourier integral, if the coefficients in (3.4) are replaced by δ-functions

$$
\begin{aligned}
\lambda &= \frac{e}{C_o} \int_{-\infty}^{\infty} d\omega \sum_{n=-\infty}^{\infty} \delta(\omega - n\omega_o) \cos \omega t \\
&= \int_{-\infty}^{\infty} d\omega \Lambda(\omega) \cos \omega t
\end{aligned}
\tag{3.5}
$$

where the frequency domain function is given by

$$
\Lambda(\omega) = \frac{e}{C_o} \sum_{n=-\infty}^{\infty} \delta(\omega - n\omega_o).
\tag{3.6}
$$

This shows that the signal is also a "comb" of equally spaced equal strength lines in the frequency domain. Pictorially the situation is shown in Fig. 1.

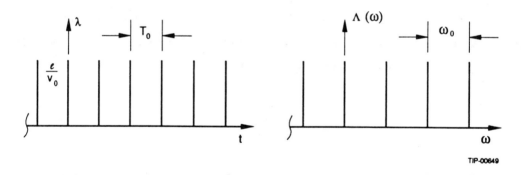

TIP-00649

Figure 1. Time domain and frequency domain spectra of a single circulating particle.

Gaussian Bunch. The line density of a bunch containing unit charge, having Gaussian profile with r.m.s. length σ_s is

$$\lambda = \frac{1}{\sqrt{2\pi}\sigma_s} e^{\frac{-s^2}{2\sigma_s^2}} = \frac{1}{\sqrt{2\pi}\sigma_s} e^{\frac{-v_0^2 t^2}{2\sigma_s^2}}$$

$$= \int_{-\infty}^{\infty} d\omega \Lambda(\omega) \cos \omega t \qquad \text{where} \qquad \Lambda(\omega) = e^{\frac{-\sigma_s^2 \omega^2}{2v_0^2}} \qquad (3.7)$$

Accounting for all beam revolutions, the time domain formula for the line charge of a bunch containing N charges e is

$$\lambda = \frac{Ne}{\sqrt{2\pi}\sigma_s} \sum_{l=-\infty}^{\infty} e^{\frac{-v_0^2(t-lT_0)^2}{2\sigma_s^2}}. \qquad (3.8)$$

This can be regarded as the convolution of distributions (3.2) and (3.7). According to a theorem of Fourier analysis, convolution in the time domain corresponds to multiplication, in the frequency domain, of the two transforms. As a result

$$\Lambda(\omega) = \frac{Ne}{C_0 v_0} e^{\frac{-\sigma_s^2 \omega^2}{2v^2}} \sum_{n=-\infty}^{\infty} \delta(\omega - n\omega_0). \qquad (3.9)$$

Pictorially the situation is shown in Fig. 2.

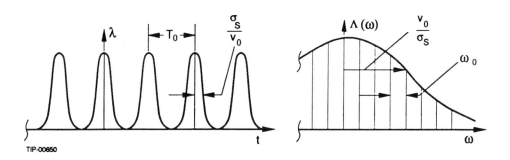

TIP-00650

Figure 2. Time domain and frequency domain spectra of a bunch of length σ_s.

Single Particle Executing Longitudinal Oscillations. Suppose the bunch of particles is executing longitudinal oscillations. Ordinarily the oscillation amplitude is smaller than the bunch length but, in the interest of simplicity, only the case of a point charge (or equivalently a sufficiently short bunch) will be illustrated. The arrival time of the particle is modulated away from its nominal value by a sinuisoidal factor oscillating at the synchrotron oscillation frequency ω_s, and with longitudinal amplitude $v_o T_s$. Substitution into (3.2) yields $\lambda = \frac{e}{v_o} \sum_{l=-\infty}^{\infty} \delta(t - lT_o - T_s \cos \omega_s t)$. This "phase modulated" expression can be expressed as a sum of harmonics of the fundamental, along with "synchrotron sidebands" that are displaced away from them by small integer multiples of the synchrotron frequency. The coefficients in this expansion are proportional to Bessel functions $J_m(n\omega_o T_s)$, where $m = 0, \pm 1, \pm 2 ...$ are labels for the sidebands, and $n = 0, \pm 1, \pm 2 ...$ are labels for the harmonics of the fundamental. Typically the "modulation depth" T_s / T_o is a very small number, so the arguments of the Bessel functions are very small compared to 1, at least for small n (i.e. low harmonics.) In that case, the leading term, with coefficient $J_o(n\omega_o T_s)$ is dominant; that makes the sidebands insignificant. At large values of n these become relatively more important. For more detail consult the first reference. Pictorially the situation is illustrated in Fig. 3.

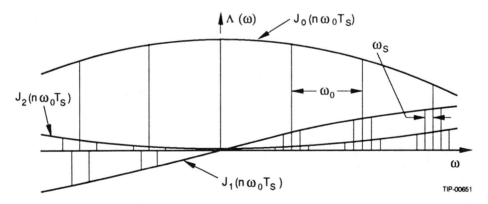

Figure 3. Frequency domain spectrum of a single particle executing longitudinal oscillations.

4. Stripline Detectors

A stripline detector in a beam tube of radius R is shown schematically in Fig. 4, which also shows an on-axis beam passing. Let the beam current be I_B within the bunch length S_B and zero elsewhere.

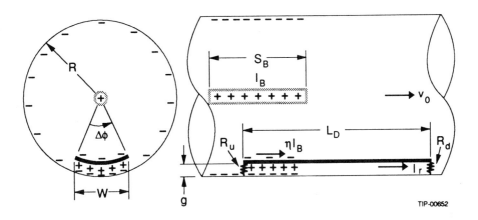

TIP-00652

Figure 4. End view and side view of a stripline detector in a cylindrical conducting chamber, as an on-axis bunch of charge passes.

If the beam is relativistic, as we assume, the electric field lines fan out isotropically at right angles to the axis. Before the bunch enters the detector the field lines terminate on charges moving along the inner surface of the vacuum chamber. While the bunch is within the detector, a fraction $\eta = \Delta\phi/(2\pi)$ of the field lines terminate on the upper surface of the electrode. Depending on the value of the upstream terminating resistor R_u, some of this charge comes from the lower surface of the electrode, and some flows through R_u. As a result of the latter, a voltage wave is launched on the transmission line formed by the electrode and the the vacuum chamber.

Some symbols to be used are:

R_u/R_d: upstream/downstream termination resistance

$L_D/w/g$: length/width/gap of stripline

η : fraction of azimuth subtended by electrode ($= \Delta\phi/(2\pi)$)

Z_o: characteristic impedance of stripline ($\simeq 377(\text{ohms})g/w$)

Right-Traveling Wave. The bunch is traveling in the positive (left-to-right) direction. While it is over R_u there is a voltage V_r (to be determined) across R_u, causing a voltage wave of that amplitude to be launched to the right. (The subscript r stands for right-traveling.) On a transmission line of characteristic impedance Z_o, a voltage wave like this is accompanied by a right-traveling current wave of amplitude

$$I_r = V_r/Z_o. \tag{4.1}$$

This current flows on the bottom surface of the electrode; an equal and opposite current flows on the chamber wall. Use of Gauss's theorem, along with the previous discussion of field-lines terminating on the electrode, shows that a current $-\eta I_B$ flows on the top surface of the electrode. Using Kirchoff's current law, the current through R_u is the sum of these two currents, and using Ohm's law, the voltage V_r is

$$V_r = (-I_r + \eta I_B)R_u. \tag{4.2}$$

From (4.1) and (4.2) one obtains

$$V_r = \eta I_B \frac{R_u Z_o}{R_u + Z_o} \equiv (R_u||Z_o)\eta I_B. \tag{4.3}$$

As a mnemonic, this can be remembered as the voltage pulse induced by a current source ηI_B driving R_u and Z_o in parallel. If the end is shorted, there is no traveling wave; if it is open, the voltage is $\eta I_B Z_o$.

To simplify the continuing discussion we assume that both the wave on the stripline and the beam particles travel at the speed of light c. That means that the front edge of the wave and the front edge of the beam bunch arrive at the downstream end simultaneously. Temporarily, we make the further, inessential, assumption that the bunch length is less than the length of the electrode. ($S_B < L_D$). The simplest case to analyse has the downstream end of the line "terminated", (i.e. $R_d = Z_o$). In that case the right-traveling wave is absorbed in R_d and there is no reflection; the wave contributes voltage V_r to the voltage appearing across R_d. More generally, there will also be a reflected, left-traveling, wave. Knowing that the reflection coefficient is $(R_d - Z_o)/(R_d + Z_o)$, the voltage appearing across R_d due to the arriving wave is

$$(1 + \frac{R_d - Z_o}{R_d + Z_o})V_r = \frac{2R_d}{R_d + Z_o}\frac{R_u Z_o}{R_u + Z_o}\eta I_B. \tag{4.4}$$

It must be remembered though, that the current flowing on the upper surface of the electrode also passes through, and contributes to the voltage across, R_d. Applying the superposition principle, this can be calculated without any reference to the just discussed arriving wave. Applying the same reasoning to the downstream end as was previously applied to the upstream end to obtain (4.3), a left-traveling voltage wave is launched with amplitude

$$V_l = -\eta I_B \frac{R_d Z_o}{R_d + Z_o}. \tag{4.5}$$

Adding (4.4) and (4.5), the voltage appearing at the downstream end, while the bunch is over it, is

$$V^{(d)} = \frac{R_d Z_o}{R_d + Z_o}\frac{R_u - Z_o}{R_u + Z_o}\eta I_B. \tag{4.6}$$

If neither end of the detector is terminated, the subsequent signals are rather complicated, consisting of multiple reflections. (In practice, a cable is attached to one or both ends in order to extract a beam position signal. It is sensible to make the cable impedance equal to the strip-line impedance, so that the line *is* terminated.) If the upstream end is terminated it can be seen from (4.6) that no signal is observed at the downstream end; the right-traveling wave from the upstream end exactly cancels the left-traveling wave from the downstream end, independent of the value of the downstream terminating resistance.

If both ends are terminated one has a remarkable directional feature; particles traveling left-to-right give a signal only on the left, and particles moving right-to-left give a signal only on the right. Accelerator "lore" at some laboratories has it that this capability is not very practical, owing to various imbalances and non-ideal effects; nevertheless directional sensitivity of 30 dB in the ratio of intensities has been obtained routinely at other laboratories.

The story is not yet complete, as it remains to calculate the signal at the left end at a time $2L_D/c$ after the right-traveling bunch first passed over it. At that time the reflection of the wave launched initially at the left end, and the wave launched at the right end, have both arrived. To simplify this, let us assume the left end is terminated.

$$V^{(u)}(\text{delayed}) = (\frac{R_u Z_o}{R_u + Z_o}\frac{R_d - Z_o}{R_d + Z_o} - \frac{R_d Z_o}{R_d + Z_o})\eta I_B = -Z_o\eta I_B/2. \qquad (4.7)$$

This is an inverted copy of the pulse observed earlier at the same point; the delay is $2L_D/c$. This combined delay and subtraction results in a bipolar signal having no net area. If the beam pulse is long compared to the electrode, the delay and subtraction is approximately the same as differentiation. For a typical value of L_D, say 7.5 cm, the effective "frequency" of the bipolar pulse is about 1000 MHz.

5. Beam Current Transformers

Passive Detection. For measuring slowly varying beam currents, transformers are often used. For this section we take the bunch length S_B to be arbitrarily long, but we assume that the current starts abruptly at $t = 0$; that is

$$I_B = I_{Bo}U(t) \tag{5.1}$$

where $U(t)$ is the unit step function, zero for negative time, one for positive time. The Laplace transform of this beam current is $\mathcal{L}\{I_B\} = I_{Bo}/s$. The basic beam current transformer pickup circuit is shown in the Fig. 5.

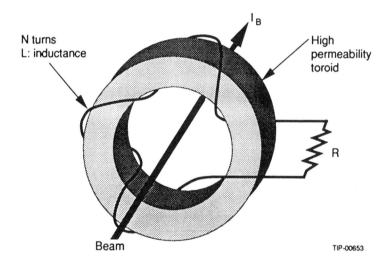

Figure 5. Beam current transformer for which the beam acts as a single turn primary current that induces a voltage across the resistor R that completes the secondary circuit.

The beam acts as a single turn primary winding. It induces a current in the secondary winding which has N turns and inductance L. The current can be measured by measuring the voltage across the resistor R. Ideally the transformer "steps down" the current by the turns ratio N, making the observed voltage $I_{Bo}R/N$. An idealized, or mid-band, equivalent circuit is shown in Fig. 6.

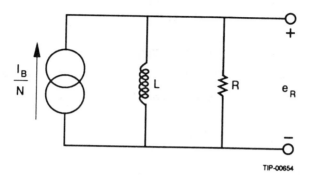

TIP-00654

Figure 6. Mid-band equivalent circuit of a beam current transformer.

In this circuit the beam current, derated to I_{Bo}/N, drives the parallel combination of the secondary inductance L and the load R. Note that in this equivalent circuit, the way the transformer has been replaced, there is no current flowing in L at $t = 0+$. Using the formalism of Laplace transforms, with $s = j\omega$, the complex impedance of the load is $RLs/(R + Ls)$. The Laplace transform of the voltage across R is

$$\mathcal{L}\{e_R\} = \frac{RLs}{R + Ls}\mathcal{L}\{I_B/N\} = \frac{I_{Bo}R}{N}\frac{1}{s + R/L}. \qquad (5.2)$$

Performing the inverse transformation yields $e_R = \frac{I_{Bo}R}{N}e^{-(R/L)t}$, as sketched in Fig. 7.

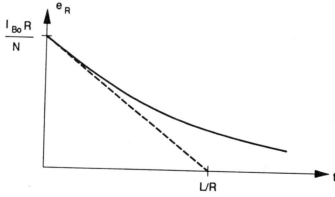

Figure 7. Response as a function of time for a beam current transformer.

For early times the "ideal" signal $I_{B_0}R/N$ is observed, but it decays away (In the circuit model any current reduction in R is reflected in an equal current increase in L.) with time constant $\tau = L/R$. It is desirable for this to be as long as possible, if slowly varying beam currents are to be measured. This can be accomplished by making L large, (which inevitably impairs the high frequency response, as will be discussed next), or by reducing R, (which reduces the output voltage, thereby reducing the signal-to-noise ratio.) It will be shown in the next section how the effective value of R can be reduced using an operational amplifier.

To account for some of the inevitable imperfections, the equivalent circuit of Fig. 8 can be used. Non-ideal or stray elements are drawn with broken lines.

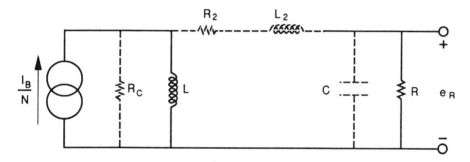

Figure 8. Broad-band circuit model for a beam current transformer; non-ideal or stray elements are drawn with broken lines.

In this equivalent circuit R_c represents transformer core losses; it is a complicated function of frequency and excitation. Leakage inductance is represented by L_2. Stray capacity is represented by C. Resistance of the secondary winding or any intentional series element is represented by R_2. This model reduces to the mid-band model when these stray elements are replaced by their ideal values, zero or infinity. With care the rise time of this device can be in the nanosecond range. A CERN ferrite core transformer, with OD/ID/thickness dimensions of 150/92/25 mm, and a 10 turn secondary, had L=0.31 mH, R=50 ohms, and L/R=6.2 μs, and rise time of 2 ns. Another 10 turn CERN model, made with high permeability tape wound core had dimensions 170/130/25 mm, L=1.6 mH, R=50 ohms, and L/R=32 μs, and rise time of 0.6 ns, clearly showing the superiority of tape wound cores, compared to ferrite.

Active-Passive Detection. It is possible to improve the low frequency response of this circuit, without impairing the high frequency response, using an operational amplifier as shown in Fig. 9.

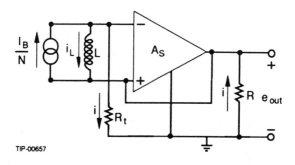

TIP-00657

Figure 9. Use of an operational amplifier in an "active-passive" detection circuit for improving the low frequency response of the beam current transformer.

The same current i flows through both resistors R and R_t, since the circuit "floats", with no current flowing to the single ground connection. The reason for having two resistors will be given later. The amplifier is a standard, high input impedance, high DC voltage gain ($A \simeq 10^6$), 6db/octave high frequency roll-off, operational amplifier, with unity gain frequency of about 100 kHz. It can be represented by a transfer function $A_s = A/(1 + s/s_o)$.

Initially neglecting the effect of the roll-off and using the "virtual ground" concept for the effect of the operational amplifier, a qualitative description of the circuit response could not be simpler. Large negative feedback holds a "virtual short circuit" between the two input terminals of the amplifier, but with no current flowing. Initially, the current flowing in L is $i_L(t = 0+) = 0$. With the voltage across L held at zero by the circuit, there can be no deviation of i_L away from zero. As a result, the initial current will continue to flow through R and R_t for all time. The fact that the amplifier does not respond at high frequency has no great effect on the output voltage. For early times, the amplifier can simply be regarded as not present in the circuit. But it can be seen that it is not being asked to respond, since the output passively jumps to its correct steady state value. This is the reason the circuit is called "active-passive". In practice, both R and R_t are the characteristic impedances of cables attached at those positions in the circuit. Also in practice, the feedback is not d.c. coupled, as shown in the figure, but rather is by transformer action via a bifilar winding on the same core; this configuration is due to Hereward at CERN.

A more quantitative analysis proceeds as follows. Three circuit equations, expressed in words are: the output voltage is the inductor voltage multiplied by the amplifier gain; the output voltage is due to current i flowing in resistance R; and the input current divides between the resistance (R and R_t in series) and the inductance. In formulas

$$\mathcal{L}\{e_{out}\} = A_s Ls \mathcal{L}\{i_L\}$$
$$= -R\mathcal{L}\{i\} \tag{5.3}$$
$$\mathcal{L}\{i\} = \mathcal{L}\{I_B/N\} - \mathcal{L}\{i_L\}$$

Solving these equations, one obtains

$$\mathcal{L}\{e_{out}\} = \frac{-RI_{Bo}/(sN)}{1 - \frac{(1+s/s_o)R}{ALs}}. \tag{5.4}$$

The early time (or high-frequency) response is obtained by allowing s to approach ∞; taking A large, the resulting formula for e_{out} is

$$e_{out} = -R(I_{Bo}/N)U(t). \tag{5.5}$$

The long-time (low frequency) response is obtained from the small-s behavior; provided the frequency is not too low, the second term in the denominator of (5.4) can still be neglected, extending the validity of (5.5) for long times.

There are real subtleties in the biasing of a high impedance circuit like this. It shares the difficulties of all DC amplifiers. It is necessary for the DC levels to be restored by periodic resetting. This is too technical for the present notes.

Incorporation of Coax Cables. It is usually necessary for the three main elements of this device, transformer, op-amp, and metering resistor, to be situated in three different places, connected by coaxial cables. The schematic of Fig. 10 shows how that can be done, compatibly with the just-completed analysis. The resistors R and R_t are chosen to equal the cable impedance. This suppresses reflections; for example those resulting from the fact that the operational amplifier presents a time varying termination to the transmission line. The only new elements C_d and R_d, present for biasing purposes, are chosen to have such a large time constant to have no effect on the previous description. With a circuit such as this, time constants on the order of a second can be achieved.

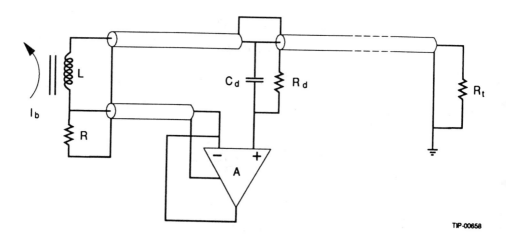

Figure 10. Layout of the elements of an "active-passive" detection circuit for a beam current transformer, showing how coaxial cables can be incorporated consistent with the analysis in the text.

6. The Direct Current Transformer.

The term "Direct Current Transformer" may be regarded as something of a contradiction in terms. Other terms employed for apparatus such as that to be described in this section are "Second Harmonic Detector" and "DC Current Transductor". Second harmonic detection is based on nonlinear circuit elements having response characteristic like that shown in Fig. 11. The idea for this is not new as the following chronological listing indicates; (this should not be regarded as historically reliable.)

Some Milestones.

1904: patent awarded

1924: Bell Telephone magnetic amplifier

1949: Q. Kern's, LBL, current transductor

1950: microgauss geological magnetometer

1960: Hewlett-Packard clip-on ammeter

1960's: high current (2000 amp) high power (0.5 Mw) power supply regulators, Princeton-Pennsylvania Accelerator

1969 to present: Unser et al., CERN current monitors

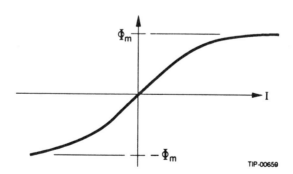

Figure 11. Anti-symmetric, nonlinear, response of the flux Φ to the current I for a high permeability toroid. This is the basic element upon which second-harmonic detection can be based.

The nonlinear element to be used here is the same toroidal core as has been discussed in the previous section. For this discussion the current I in a winding can be regarded as the input variable, and the flux Φ in the core can be regarded as the output variable. An essential feature is that the toroid "saturates" at flux level Φ_m. Ideally the input-to-output transfer relation has the form shown in Fig. 11. Any device exhibiting nonlinear input-to-output transfer response that is anti-symmetric for negative and positive inputs can be used for second harmonic sensing. The most common system of this sort, and the only one to be discussed here is based on a toroid of high magnetic permeability, much like the beam current transformer cores described above.

Mathematically, the relationship can be written

$$\Phi = a_1 I + a_3 I^3 + a_5 I^5 + \ldots \tag{6.1}$$

where only odd powers appear. If a sinuisoidal current is applied

$$I = I_1 \cos \omega t, \tag{6.2}$$

the output will take the form

$$\Phi = \Phi_1 \cos \omega t + \Phi_3 \cos 3\omega t + \ldots, \tag{6.3}$$

where, either by symmetry, or by explicit use of trigonometric identities, it can be seen that only odd harmonics appear. In particular, there is no second harmonic. But suppose that the input has a (probably small) DC offset I_0, (which is the quantity the device is intended to measure), and is given by

$$I = I_0 + I_1 \cos \omega t. \tag{6.4}$$

In that case the output is given by

$$\begin{aligned}
\Phi &= a_1(I_0 + I_1 \cos \omega t) + a_3(I_0 + I_1 \cos \omega t)^3 + \ldots \\
&= \ldots + 3 a_3 I_0 I_1 \cos^2 \omega t + \ldots \\
&= \ldots + \frac{3}{2} a_3 I_0 I_1 \cos 2\omega t + \ldots \quad .
\end{aligned} \tag{6.5}$$

The essential feature here is a term proportional to the DC offset I_0, oscillating at the second harmonic frequency. Devices based on this principle detect the presence of this term by detecting any second harmonic signal synchronous with the input signal, and using that signal in a closed feedback loop to null out the second harmonic component of the output. In other words, an intentional offset current I_f is added (using a different winding) and the result of the "measurement" is $-I_f(\text{null})$, the value of I_f that suppresses the second harmonic. From (6.5) it can be seen that the sensitivity is proportional to the amplitude I_1, of the input drive. In a magnetic circuit, I_1 is chosen large enough to drive the core into saturation in both polarities. Clearly proper circuit operation depends on the saturation characteristics being symmetric. Hysteresis, for example, impairs the operation, so special magnetic materials are called for.

To understand circuits based on this principle it is helpful to understand first how switching cores can be used for rectification. The circuit of Fig. 12 illustrates this. Waveforms in this circuit are plotted in Fig. 13.

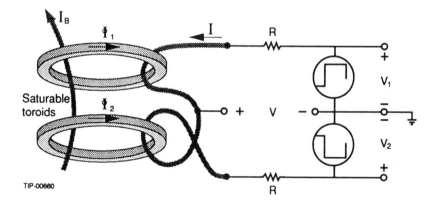

Figure 12. Circuit to generate a "rectified" voltage V. Symmetric square wave signals are applied to the series windings of two toroids linked by a d.c. current I_B.

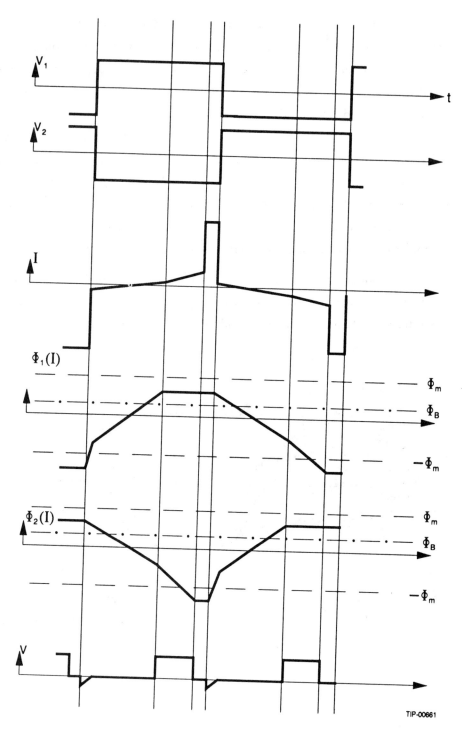

Figure 13. Time dependence of various waveforms in the circuit of Fig. 12.

TIP-00661

The d.c. current I_B causes equal flux Φ_B in the two toroids. While either toroid is unsaturated it presents a large inductance and limits the current I to a small value. The equal, but opposite, square-wave voltages V_1 and V_2 have great enough amplitudes to drive both cores into saturation temporarily; that accounts for the narrow spikes in I. The current in these spikes is limited by the series resistors R. Because of the bias flux Φ_B, due to I_B, there is an interval during which one toroid is saturated, while the other is not. During this interval there is large voltage across one toroid and none across the other; that accounts for the broad pulses of V, shown in Fig. 13. The solid curves labelled $\Phi_1(I)$ and $\Phi_2(I)$ show the time dependence of the fluxes caused by the current I. Saturation for the first core occurs when $\Phi_1(I) + \Phi_B = \Phi_m$. Starting then the voltage pulse V lasts during the interval of time during which the flux $\Phi_2(I)$ increases negatively (more rapidly than before because of the doubled e.m.f. applied to it) until $\Phi_2(I) + \Phi_B = -\Phi_m$.

It is up to the reader to study the polarities, to persuade him or herself that the fundamental frequency of the current I is the same as that of the square-wave drive, while the voltage V, as well as having a d.c. or "rectified" component, varies at twice the drive frequency. A circuit exploiting this behavior senses this second harmonic component and nulls it out as has been explained previously.

The schematic of Fig. 14, due to Unser, shows a circuit designed to measure the current I flowing in series through identical windings on two identical "switching cores". A symmetric square wave voltage V_ϕ is applied symetrically through the balanced transformer on the right. The windings on the switching cores are such that one core saturates during one half cycle, and the other saturates during the other half cycle. The 500 Hz oscillator serves the dual purpose of synchronous sensing of the second harmonic signal, and after frequency division by two and amplification, of switching the cores.

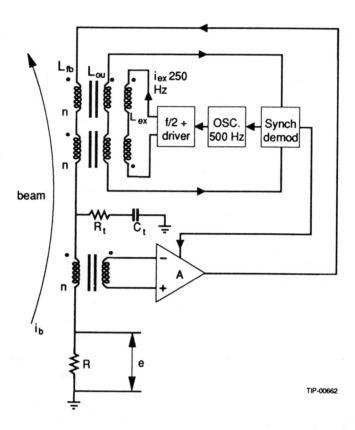

Figure 14. Practical circuit, due to Unser, employing the direct current transformer principle to obtain time constants of the order of a day.

7. References

Much of the content of these lectures has been extracted from the following papers.

Raphael Littauer, *Beam Instrumentation*, in Physics of High Energy Accelerators, AIP Proceedings, No. 105, 869-953 (1983).

Klaus Unser, *A Toroidal D.C. Beam Current Transformer With High Resolution*, CERN-ISR-OP/81-14, and 1981 Particle Accelerator Conference, Washington, 1981. Also. *Beam Current Transformer With DC to 200 MZ Range*, CERN-ISR-CO/69-6. Also. *Toroidal AC and DC Current Transformers for Beam Intensity Measurements*, Atomkernenergie, Kerntechnik, Vol. 47 (1985).

Anonymous notes in my possession, written, it seems, by someone who was once at PPA. *Second Harmonic Detector*.

BEAM POSITION MONITORING

Robert E. Shafer

Los Alamos National Laboratory, Los Alamos, NM 87545

1. INTRODUCTION

The purpose of this paper is to review the topic of beam position monitoring as it applies to charged-particle beams in accelerators and beamlines. As this workshop is primarily to familiarize engineers with the subject, the emphasis will be on the engineering aspects of beam position measurement rather than on pedantic derivations of equations. The intent is not to present many specific solutions to specific problems, but to provide general guidelines on which specific designs can be based, with occasional examples. Some math must be included, however, to understand how pickup electrodes respond to beams and how various circuits process signals. The emphasis will be on calculating frequency dependence, power levels and beam-displacement sensitivities, and reviewing the advantages and disadvantages of a variety of electrode designs and signal-processing methods.

Beam instrumentation includes not only the electronics necessary to measure the properties of beams, but also the transducers that convert the beam signals into electrical signals. Processing electrical signals is a very well understood engineering field. The characteristics of the transducers that generate the electrical signals are not well documented, however. There is quite a bit of physics in beam transducers, and because the beam currents are not confined to wires, engineers sometimes have difficulty grasping the basic concepts in how particle beams couple to beam diagnostic devices and generate electrical signals. Therefore, this paper will concentrate on the beam-coupling mechanism, but will also discuss some of the signal-processing electronics.

First, the basic methods of noninterceptive beam position monitoring will be discussed. Then the basic characteristics of beams and beam position measuring systems will be reviewed. This is followed by several sections that discuss the signals from a variety of pickup electrode geometries. Several sections are then devoted to signal processing methods. Finally, several sections discuss the system aspects of beam position monitoring.

There have been several comprehensive survey articles on beam diagnostics stressing primarily the physics aspects of the subject.[1-6] These articles cover the broad range of particle beam diagnostics and are not restricted to position measurement.

2. BASIC METHODS OF BEAM POSITION MONITORING

The most common method of monitoring the position of a charged-particle beam is to couple to the electromagnetic field of the beam. The beam is a current, and it is therefore accompanied by both a magnetic field and an electric field. In the limit of very high beam energy, the fields are pure transverse electric and magnetic (TEM). If the beam is displaced from the center of a hollow conducting enclosure, the magnetic and electric fields are modified accordingly. Detailed knowledge of how the magnetic and electric fields depend on the beam position allow accurate determination of the beam position.

Pickup electrodes, in general, cannot sense dc electric or magnetic fields (there are exceptions to this, of course, such as Hall probes and flux-gate magnetometers). The signals are induced by a time-varying component of the beam signal, usually beam current modulation. The carrier for the beam position information is the frequency (and harmonics) of the periodic beam bunches for a continuous train of bunches or the derivative of the instantaneous beam current for single bunches.

The conventional beam position pickup is a pair of electrodes (or two pair, if two beam position coordinates are being measured) on which the signals are induced. The ratio of the amplitudes of the induced signals at the carrier frequency is uniquely related to the beam position. Because the position information is contained in the amplitude variation of these signals, the information actually appears as AM (amplitude modulation) sidebands of the carrier. In synchrotrons, where the strong focusing forces cause the betatron oscillation frequency to be many times the revolution frequency, the sidebands are substantially displaced in frequency from the carrier signal.[3,4]

A variant of the standard multielectrode beam position monitor is the wall-current monitor,[7] which often is a gap in the beam pipe with a ceramic insert to maintain vacuum. A resistive path, often composed of many resistors in parallel, is connected across the gap to carry the wall currents. The azimuthal distribution of wall currents, determined by measuring the voltage drop across the resistors, can be used as a measurement of the beam position.

Another nearly noninterceptive method of measuring position is to intercept the beam with objects such as thin wires and to measure the response of the wire (induced currents) or ionizing radiation produced by the beam hitting the wire as the wire is moved through the beam.[8] The wire, usually a low-Z material, may be either stepped through the beam or pushed through at a high velocity.[8] In either case, some of the beam is lost or scattered out of the incident beam, causing some degradation of the beam quality. In addition, the wire is heated by the beam and, in many cases, is damaged by being overheated.

Still another method is to cause the beam itself to radiate energy. High-energy electrons, when they pass through large magnetic fields, will radiate visible, uv light or x-rays in narrow forward cones (synchrotron radiation). Detectors designed to intercept this synchrotron radiation can be used to determine the position of the source (i.e., the beam) in the magnetic field.[9] This general technique is used in electron synchrotrons (synchrotron light sources) and in devices such as wigglers and undulators, which are specifically designed to make the electrons radiate.

Residual gas fluorescence[10] or ionization[11] can be used to determine beam position. In the former, video-image processing of the visible light is used to determine the beam profile and position. In the latter, an electric field perpendicular to the beam is used to accelerate the liberated electrons or ions to a detector.

Finally, it is possible to probe the beam with another beam, and monitor the radiation produced by the two interacting beams. At the Stanford Linear Accelerator Center, for example, two very high energy beams (one an electron beam and the other a positron beam), are directed toward each other in the collision (final focus) area of the Stanford Linear Collider (SLC). [12,13] The magnetic fields of each beam cause the other beam to be deflected and radiate (beamstrahlung). Measurement of the radiation yield as a function of the relative position of the two beams leads to a measurement of the relative beam positions.

In the case of negative-ion beams such as H^- (a proton with two electrons attached), a thin visible laser beam (photon energy about 2 eV) can be used to detach one of the electrons[14] (the binding energy of the "extra" electron is about 0.75 eV). The result is that a narrow beam of neutral hydrogen atoms is produced. By monitoring the yield of the neutral hydrogen as a function of the position of the laser beam, the beam position can be determined. Laser beams may also be scattered off high-intensity electron beams as a probe.[15]

There are many ways of determining the beam position in a nearly noninterceptive way. The best one is determined by the type of beam being monitored, the conditions under which the monitoring is being done, and, finally, what specific measurements are required. Because the most common noninterceptive method is measuring the electric and magnetic fields, the remainder of the paper will discuss this method.

3.   CHARACTERISTICS OF BEAMS AND POSITION MEASURING SYSTEMS

The purpose of this section is to discuss some of the beam and beam position measuring system characteristics that need to be considered when such a system is being designed. It is important to understand the range of the beam parameters to be expected and the requirements of the beam position monitoring system before undertaking a detailed design of the system. It is nearly always true that some compromises must be made in order to go from an ideal system design to a realizable one. These compromises may be due to time constraints, funding, space, or manpower resources. Hence, it is important to have a thorough understanding of how each parameter affects, or is affected by, the system design. With this understanding, it is usually possible to design a system that is simple, yet does not compromise the quality of the measurements.

Accuracy is the ability to determine the position of the beam relative to the device being used for measuring the beam position. This is limited by some combination of mechanical alignment errors, mechanical tolerances in the beam detection device, calibration errors in the electronics, attenuation and reflections in the cables connecting the pickup to the electronics, electromagnetic interference, and circuit noise (noise figure of the electronics). Signal processing introduces additional inaccuracies such as granularity (least-significant-bit [LSB] errors) that is due to analog-to-digital conversion.

Resolution differs from accuracy in that it refers to the ability to measure small displacements of the beam, as opposed to its absolute position. Typically, the resolution of a system is much better than the accuracy. In many cases, good resolution is much more important than good accuracy. For example, it is often adequate to know the absolute beam position to a fraction of a millimeter, even though the beam motion (jitter) needs to be known to a few micrometers. In high-energy collider operation, for example, it is much more important to know the relative position of the two beams than to know the absolute position of either.

Bandwidth refers to the frequency range over which beam position can be measured. In some cases, a beam may have a fast transverse motion (jitter) that needs to be identified. In another case, the beam pulse may be very short (a picosecond, nanosecond, or a microsecond for example), and the measuring system must be able to acquire data in this time interval (acquisition bandwidth). Closely related is real − time bandwidth, which is the ability to generate a real-time analog signal proportional to the beam position in a limited time. This response is necessary if the signal is to be used in real-time, closed-loop control applications.

Beam current usually refers to the average (dc) beam current averaged over the microscopic bunch structure, but can also be used to refer to the instantaneous (intrabunch) beam current and to other temporal averages. For single bunches, the number of particles per bunch is often used as a measure of beam current. Closely related is beam intensity, which usually refers to the amplitude of a particular frequency harmonic of the beam-bunching frequency. Beam intensity differs from beam current in that intensity is a frequency-domain quantity, while current is a temporal-domain quantity.

Dynamic range refers to the range of beam intensities (or current or charge) over which the diagnostic system must respond. Often large dynamic range response is achieved by gain switching, but in addition, special methods of signal processing can provide a large dynamic range response and eliminate the need for gain switching.

Signal − to − noise ratio refers to the relative power levels of the wanted signal to unwanted noise. Noise may be true thermal noise, amplifier noise (noise figure), electromagnetic noise (EMI) such as silicon-controlled-rectifier (SCR) noise, or radio-frequency interference (RFI), which may be the same frequency as the beam position signals. In this application, shot noise (sometimes called Schottky noise) from the beam itself is actually a signal, because it can be used to determine the beam position. Signal-to-noise ratios place limits on the ultimate resolution of the system.

Beam bunching refers to the temporal characteristics of the beam current modulation. Usually the beam is in the form of short bunches with the same period as, or a multiple of, the period of the rf system being used to accelerate it. For example, at the Los Alamos Meson Physics Facility (LAMPF), the bunch period is about 5 ns (201.25 MHz), while the rf period is 1.25 ns (805 MHz). The bunch length is usually quite short relative to the period and at some facilities is less than 30 ps. The bunch shape can be temporally symmetric such as Gaussian, parabolic, or cosine-squared, among others. It can also be nonsymmetric. The beam-bunching factor is typically the ratio of the bunching period to the beam bunch full length at half maximum (F L HM). Typically, this factor can be 10 or 20, often higher. This temporal profile creates many harmonics of the bunching frequency in the induced signals on the pickup electrodes. Beam bunching can change with time because of momentum spread in a nonisochronous beam-transport system, synchrotron oscillations in rf buckets, or by allowing a space-charge-dominated beam to coast in a beamline without longitudinal focusing forces.

4. BEAM CURRENT MODULATION IN THE TIME AND FREQUENCY DOMAINS

Beam bunches can have many shapes. Regardless of what the specific shape is, the beam-bunching frequency usually provides the carrier signal that is used for detecting the beam position. Because it is possible to make measurements in either the time or frequency domain, it is important to understand the interrelations between the two cases. A Gaussian bunch shape is used in the following calculations, although other shapes could just as easily have been used.

Consider a Gaussian-shaped beam bunch containing N particles of charge e in a bunch of rms temporal length σ (in time units) and with a bunching period T. The instantaneous beam current of a single bunch is given by

$$I_b(t) = \frac{eN}{\sqrt{2\pi}\sigma}\exp\left[\frac{-t^2}{2\sigma^2}\right] \quad . \tag{4.1}$$

This is normalized so that the bunch area is the total charge eN independent of the rms bunch length σ. Assuming that the bunch is symmetric in time, centered at $t = 0$, and is in a pulse train with bunch spacing T, we can expand this in a cosine series with $\omega_0 = 2\pi/T$:

$$I_b(t) = \frac{eN}{T} + \sum_{m=1}^{\infty} I_m \cos\left(m\omega_0 t\right) \quad , \tag{4.2}$$

where

$$I_m = \frac{2eN}{T}\exp\left[\frac{-m^2\omega_0^2\sigma^2}{2}\right] \quad . \tag{4.3}$$

This may be rewritten

$$I_b(t) = \langle I_b \rangle + 2\langle I_b \rangle \sum_{m=1}^{\infty} A_m \cos\left(m\omega_0 t\right) \quad , \tag{4.4}$$

where the average (dc) beam current is

$$\langle I_b \rangle = \frac{eN}{T} \quad , \tag{4.5}$$

and the harmonic m amplitude factor is

$$A_m = \exp\left[\frac{-m^2 \omega_0^2 \sigma^2}{2}\right] \quad . \tag{4.6}$$

The Fourier cosine series expansion of the beam current in Eq. (4.4) includes a dc component as well as many harmonics of the bunching frequency. The amplitude (intensity) of the various Fourier harmonics is determined by the factor A_m, which always approaches 1 for small harmonic numbers, regardless of the specific bunch shape. As the bunch shape approaches a δ function, the amplitude factor A_m approaches 1 for all harmonics. The peak amplitude of the low harmonics of the bunching frequency is about twice the dc current. Table 4.1 gives the amplitude factor A_m for a variety of pulse shapes.

Table 4.1. Amplitude factor A_m for a variety of bunch shapes. All expressions are normalized so that $A_m \rightarrow 1$ as the bunch length $\rightarrow 0$. In the table, T is the bunch spacing, W = full width at base of bunch, and $\omega_0 = 2\pi/T$.

Bunch Shape	A_m	Comment
δ function	1	For all harmonics
Gaussian	$\exp\left(\frac{-m^2 \omega_0^2 \sigma^2}{2}\right)$	σ is rms bunch length
Parabolic	$3\left(\frac{\sin\alpha}{\alpha^3} - \frac{\cos\alpha}{\alpha^2}\right)$	$\alpha = m\pi W/T$
$(\text{cosine})^2$	$\frac{\sin(\alpha-2)\pi/2}{(\alpha-2)\pi} + \frac{\sin\alpha\pi/2}{\alpha\pi/2} + \frac{\sin(\alpha+2)\pi/2}{(\alpha+2)\pi}$	$\alpha = 2mW/T$
Triangular	$\frac{2}{\alpha^2} - \frac{2\cos\alpha}{\alpha^2}$	$\alpha = m\pi W/T$
Square	$\frac{\sin\alpha}{\alpha}$	$\alpha = m\pi W/T$

It should be noted that many bunch shapes can have zero values of A_m for certain harmonics m of the bunching frequency, depending on the bunch length. If a beam position system is being designed to operate at a harmonic of the bunching frequency, this must be taken into account. If the bunch shape is variable, then the amplitude factor may vary and may even go to zero, depending on the specific bunch shape and length.

In summary, the currents associated with periodically spaced beam bunch may be considered either in the time domain or the frequency domain. Generally, if the signal processing is performed at harmonic m =1 in the frequency domain, the amplitude factor A_1 is nearly 1, and the rms beam intensity at this frequency is $\sqrt{2}$ times the dc current.

An interesting alternate method to using the rf bunching modulation is the beam current modulation scheme being planned at CEBAF. A 1-μA rms beam current modulation at 10 MHz is placed on the 200-μA cw electron beam[16] (rf frequency is 1497 MHz). Because the 10-MHz modulation can be turned on for less than one revolution around the recirculating linac (period about 4.2 μs), it is possible to measure the position of individual orbits while the machine is in operation. There are two other advantages to this method: First, the instrumentation is less expensive at 10 MHz, and second, the modulation scheme can be used without adversely affecting running experiments.

If a beam is centered in a circular, conducting beam pipe of radius b and has a velocity $v_b = \beta_b c$ (where c is the speed of light), then there is an electromagnetic field accompanying the beam and an equal magnitude, opposite charge, uniformly-distributed beam current density on the inner wall of the beam pipe. The field inside the beam pipe looks (nearly) like a transverse-electric-magnetic (TEM) wave propagating down the beam pipe at the beam velocity (this is exact only for $\beta_b = 1$). Beam position detectors sense these fields (or, equivalently, the corresponding wall current) and determine the beam position based on the relative amplitudes of the induced signals in two or more pickup electrodes. The instantaneous Fourier harmonic amplitudes of the wall currents (integrated over 2π), in this case, are the same as those for the beam itself. The wall current density is then, for a beam pipe with infinite conductivity, simply the beam current divided by the beam-pipe circumference:

$$i_w(t) = \frac{-I_b(t)}{2\pi b} \tag{4.7}$$

Some authors like to differentiate between pickups that detect the TEM fields and those that sense the wall currents. There is no difference. For a beam current $I_b(t)$ in the center of a conducting beam pipe of radius b, the azimuthal magnetic field accompanying it is $H_\theta(r,t) = I_b(t)/2\pi r$. Because $[\text{curl } \mathbf{H}(t)]_z = J_z(t)$, the discontinuity of $H_\theta(r,t)$ at $r = b$ requires that $J_z(b,t) = H_\theta(b,t)$, as long as the magnetic fields associated with the beam are confined to the region inside the beam pipe. For this reason, we can consider either the TEM wave or the wall current density as the excitation signal. For an rf-modulated cw beam in a metallic beam pipe, the magnetic field associated with the dc component of Eq. (4.4) will eventually appear outside the beam pipe, and the wall currents will then include only the ac components.

Circular beam pipes are the most common shape, and so the rest of this paper will deal exclusively with circular geometry. Other geometries for beam pipes and beam pickup electrodes include rectangular, diamond, and elliptical, among others. All the calculations carried out in this paper can be done for these other geometries, with similar results.

5. SIGNALS FROM OFF-CENTER BEAMS

In the previous section, we considered a centered beam in a circular beam pipe. We now investigate what happens to the wall currents when the beam is displaced from the center.

LaPlace's equation can be solved in two dimensions to find the wall current density for a pencil beam current $I_b(t)$ at position r, θ inside a grounded, circular, conducting beam pipe of radius b.[17] The wall current density i_w at b, ϕ_w is then

$$i_w(b, \phi_w, t) = \frac{-I_b(t)}{2\pi b}\left[1 + 2\sum_{n=1}^{\infty}\left(\frac{r}{b}\right)^n \cos\left[n(\phi_w - \theta)\right]\right] . \qquad (5.1)$$

An alternate way to obtain a solution is to use the method of images. In this case, the location of an image pencil beam is found such that the potential everywhere on the circle corresponding to the beam-pipe location (without the beam pipe) is zero. The wall current is then calculated using the differential form of Gauss's Law ($\text{div } E = \rho/\epsilon_0$). The resultant expression for the wall current density i_w at b, ϕ_w is [18]

$$i_w(b, \phi_w, t) = \frac{-I_b(t)}{2\pi b}\left[\frac{b^2 - r^2}{b^2 + r^2 - 2br \cos\left(\phi_w - \theta\right)}\right] . \qquad (5.2)$$

This closed-form expression, which is equivalent to the infinite series form in Eq. (5.1), is sometimes easier to deal with than the infinite series. However, when the expression must be integrated, the infinite series is often the preferred form. Note that the infinite series is of the form $r^n \cos n\theta$, indicative of solutions in cylindrical geometry.

If two electrodes (L and R for left and right) of angular width ϕ are placed at $0°$ and $180°$, as shown in Fig. 5.1, the resultant currents flowing parallel to the beam on the inside surface of these electrodes are (assuming they are grounded and also at radius b)

$$I_R(t) = \frac{-I_b(t)\phi}{2\pi}\left\{1 + \frac{4}{\phi}\sum_{n=1}^{\infty}\frac{1}{n}\left(\frac{r}{b}\right)^n \cos\left(n\theta\right) \sin\left(\frac{n\phi}{2}\right)\right\} , \qquad (5.3)$$

and

$$I_L(t) = \frac{-I_b(t)\phi}{2\pi}\left\{1 + \frac{4}{\phi}\sum_{n=1}^{\infty}\frac{1}{n}\left(\frac{r}{b}\right)^n \cos\left(n\theta\right) \sin\left[n\left(\pi + \frac{\phi}{2}\right)\right]\right\} . \qquad (5.4)$$

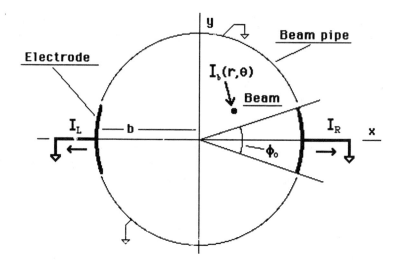

Fig. 5.1. Cross section of beam position monitor pickup model used for calculations.

$$\frac{R-L}{R+L} = \frac{4 \sin{(\phi/2)}}{\phi}\frac{x}{b} + \text{ higher order terms } . \tag{5.5}$$

A more linear (in x) approximation in cylindrical geometry is to write the ratio of R/L in decibels (i.e., logarithmic form):

$$20 \, \text{Log}_{10}\left(\frac{R}{L}\right) = xS_x = \frac{160}{Ln\ 10}\frac{\sin{(\phi/2)}}{\phi}\frac{x}{b} + \text{ higher order terms } . \tag{5.6}$$

As an example, consider two opposing 36° ($\phi = 0.63$) electrodes in a 27.5-mm radius beam pipe. The calculated sensitivity S_x is 1.25 dB per millimeter of displacement. This sensitivity, multiplied by the displacement x, gives the decibel difference of the two signals. If the displacement is 10 mm, for example, the decibel difference is about 12.5 dB, corresponding to an amplitude ratio of 4.2 to 1.

In actuality, the electrodes are neither grounded nor at the same radius as the beam pipe. In practice, the best empirical results are obtained if the radius used in Eq. (5.6) is the average of the electrode and the beam pipe radius. As an example, consider the Deutches Electronen Synchrotron Laboratory (DESY) directional coupler beam position pickup,[19] which has electrode dimensions corresponding to the above example and a ground plane radius of 32.5 mm.

The calculated sensitivity for a 30-mm effective-radius aperture is 1.14 dB per mm, which is close to the measured sensitivity of 1.18 dB per mm.

The actual measured response of the DESY pickup is a nonlinear function of displacement. Figure 5.2 shows the calculated response for a displaced beam in this geometry, using the decibel ratio of Eqs. (5.3) and (5.4). This curve agrees very well with the actual response.[19]

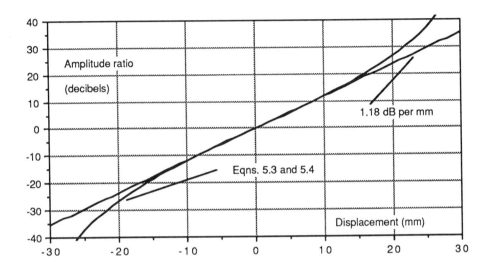

Fig. 5.2. Amplitude ratio response for a beam position monitor with a 60-mm-diam aperture and 36° electrodes. The calculated response using Eqs. (5.3) and (5.4) agrees closely with actual measurements.

Figure 5.3 shows the calculated integral linearity S_x (output amplitude divided by the x displacement) for this pickup for a vertically centered $(y = 0)$, horizontally displaced beam and also for a beam displaced vertically by $y = \pm 10$ mm, using Eqs. (5.3) and (5.4). Due to the y-plane dependence, the variation of S_x inside a 20-mm-diam circle is about $\pm 7\%$.

In general, the pickup displacement sensitivity S_x is dependent on both the in-plane and orthogonal-plane displacements x and y. The in-plane nonlinearity can be corrected after the measurement using a software algorithm or look-up table. The orthogonal plane nonlinearity cannot be corrected, however, unless the y position is also measured.

Special electrode geometries can minimize this effect. As an example, Fig. 5.4 shows the calculated response for the above pickup with the electrode width increased to 75°. In this case, the y-plane nonlinearity is nearly zero, and S_x is flat within about $\pm 2\%$ inside a 20-mm-diam circle.

Both Eqs. (5.5) and (5.6) for the electrode response to a beam displacement have higher-order terms and therefore are nonlinear. The nature of the

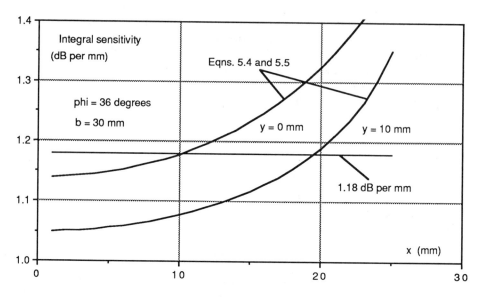

Fig. 5.3. Integral linearity plot for the pickup in Fig. 5.2 (60-mm-diam aperture with 36° electrodes). Note that the sensitivity for a vertically displaced beam ($y = \pm 10$ mm) is about 10% lower than for a vertically centered beam.

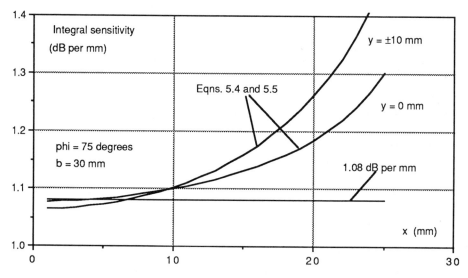

Fig. 5.4. Integral linearity plot for a pickup with a 60-mm-diam aperture and 75° electrodes. Note that the sensitivity for a vertically displaced beam ($y = \pm 10$ mm) is nearly the same as for a vertically centered beam.

nonlinearities are different for the two expressions, however, because Eq. (5.5) represents the normalized difference and Eq. (5.6) represents the logarithmic ratio. It is possible to design an electrode shape that is linear in the normalized difference, and this design will be discussed in a later section.

6. ELECTROSTATIC PICKUP ELECTRODES

We first consider the response of electrostatic (sometimes called capacitive) pickups. We consider specifically two opposing electrodes of length ℓ and azimuthal width ϕ in a beam pipe of radius b. If the current of a centered pencil beam is $I_b(t)$ and the beam has velocity $v_b = \beta_b c$, the charge density of the beam is

$$q_b(t) = \frac{I_b(t)}{\beta_b c} \ .$$
(6.1)

Equal magnitude, opposite-polarity charge appears on the inside surface of the electrodes. For an electrode of length ℓ and azimuthal width ϕ, this charge is

$$Q_s(t) = \frac{-\phi\ell}{2\pi} \frac{I_b(t)}{\beta_b c} \ .$$
(6.2)

Assuming there is capacitance between the electrode and ground given by C, the signal current flowing onto the capacitance is equal to the time derivative of the charge on the electrode:

$$i_s(t) = \frac{-dQ_s(t)}{dt} = \frac{\phi\ell}{2\pi} \frac{1}{\beta_b c} \frac{dI_b(t)}{dt} \ .$$
(6.3)

Note that there is no dc component of charge on the capacitance. The capacitance integrates this current, yielding an output voltage

$$V_c(t) = \frac{\phi\ell}{2\pi C} \frac{I_b(t)}{\beta_b c} - V_0 \ ,$$
(6.4)

where V_0 is a constant of integration.

This capacitance may be directly between the electrode and the beam pipe, or it may be added externally. The equivalent circuit is shown in Fig. 6.1. It is important to note that the signal source is a current source, and that there is some inter-electrode capacitance. In addition, there is a bleeder resistor used to prevent excessive charge build-up in the circuit. The bleeder resistor causes the average voltage on the capacitor to be zero. Usually in electrostatic pickup circuits the shunt capacitance is the dominant conductance path at the important frequencies, and the voltage across it then represents the beam bunch temporal profile, as is seen in Eq. (6.4).

As an example, consider an electrostatic pickup electrode in the Proton Storage Ring (PSR) at the Los Alamos Meson Physics Facility (LAMPF). With a bunched 20-A peak current of 800-MeV protons ($\beta_b = 0.84$), an electrode of

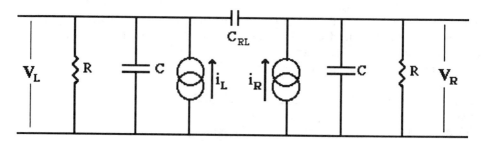

Fig. 6.1. Equivalent circuit for an electrostatic pickup. The signal sources are current generators with amplitudes specified in Eq. (6.3). Note that there is also an interelectrode coupling capacitance.

120° width and 10-cm length, and a capacitance-to-ground of 2 nF, the peak-to-peak output voltage is 1.3 V.

Usually these electrodes are coupled to the electronics by a short piece of transmission line. If neither end is properly terminated, then it is possible to excite standing waves in the cable. For this reason, special effort must be taken to damp the standing waves. The problem is most serious if the electronics is remote from the beam line, either because of the desire to have it outside the interlocked area, or to prevent radiation damage. Sometimes rad-hard electronics (in the form of a cathode follower[20]) can overcome the radiation problem.

7. LINEAR RESPONSE PICKUP ELECTRODE DESIGN

We now consider a hollow tube with radius b and length $2L$ inside a grounded beam pipe. If the tube is cut diagonally to make two electrodes as shown in Fig. 7.1, the response to beam displacement is linear. This can be seen as follows.

If a beam of charge density q_b is displaced an amount r, θ from the axis of a cylinder whose length is given by $L(\phi) = L(1 + \cos \phi)$, the total charge on the inner surface of the cylinder is

$$Q_s = -q_b L \int_0^{2\pi} \frac{(1 + \cos \phi)(b^2 - r^2)}{b^2 + r^2 - 2br \cos (\phi - \theta)} d\phi \ . \tag{7.1}$$

Upon integration (with some difficulty), this becomes

$$Q_s = -q_b L \left(1 + \frac{r \cos \theta}{b}\right) = -q_b L \left(1 + \frac{x}{b}\right) , \tag{7.2}$$

which is linear in the beam displacement. The dispacement sensitivity S_x is then given by

$$\frac{R - L}{R + L} = \frac{x}{b} , \tag{7.3}$$

which is only half the sensitivity of the electrodes in Eq. (5.5).

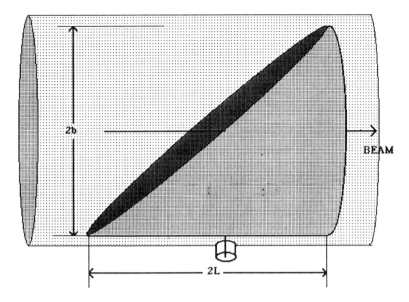

Fig. 7.1. Side view of a diagonally cut circular cross-section electrode that produces a linear response to a displaced beam. The displacement sensitivity (in decibels per unit displacement) is about 50% of a directional coupler with similar aperture.

8. BUTTON ELECTRODES

Button electrodes,[21] in common use around electron synchrotrons and storage rings, are a variant of the electrostatic electrode. The electrodes are very small and usually circular. Unlike the electrostatic electrode, however, the transmission line carrying the output signal to the electronics is terminated in its characteristic impedance, thus giving a differentiated signal. Their equivalent circuit is essentially Fig. 6.1, but with R being the terminating resistance and C being very small. Using a radius b for the beam-pipe half-aperture, the voltage onto a termination of R Ω is, using Eq. (6.3):

$$V_B(t) = R\, i_s(t) = \frac{\phi \ell R}{2\pi \beta_b c} \frac{dI_b(t)}{dt} \quad . \tag{8.1}$$

For circular elecrodes, the factor $\phi \ell$ should be set equal to the electrode area divided by b.

It is useful to combine Eq. (8.1) with Eq. (4.1) to calculate the button-electrode response to a single Gaussian beam bunch containing N particles (for $\sigma > \ell/\beta_b c$):

$$V_B(t) = \frac{-eN}{(2\pi)^{3/2}} \frac{\phi \ell R}{\beta_b c} \frac{t}{\sigma^3} \exp \left[\frac{-t^2}{2\sigma^2} \right] . \qquad (8.2)$$

The peak output voltage of this bipolar doublet, which occurs at about $t = \pm \sigma$, is then given by

$$V_{\text{peak}} = \frac{eN}{(2\pi)^{3/2}} \frac{\phi \ell R}{\beta_b c} \frac{e^{-1/2}}{\sigma^2} . \qquad (8.3)$$

This peak amplitude varies inversely as the square of the beam bunch length. For this reason, button electrodes are most useful around machines with very short bunches (i.e., electron accelerators and storage rings) and are not normally used around proton machines, which typically have longer bunch lengths (an exception to this is the PETRA ring at DESY, which will use button electrodes for monitoring both electron and proton beams).

If we use Eq. (8.1) in combination with the Fourier series expression as shown in Eq. (4.4), the rms output voltage for a beam intensity signal at frequency $\omega/2\pi$ is (where $A(w)$ is the amplitude factor Eq. (4.6) at frequency w)

$$V_B(\omega) = \frac{\sqrt{2}\phi R}{\pi} \langle I_b \rangle A(\omega) \frac{\omega \ell}{2\beta_b c} . \qquad (8.4)$$

This signal amplitude rises linearly with frequency for a given average beam current.

9. RESONANCES IN PICKUP ELECTRODES

We now consider a simple electrostatic electrode inside the beam pipe, as shown in Fig. 9.1. Assuming the only external connection to the electrode is at the center, both the upstream and downstream ends of the electrode are open (i.e., unterminated). Because the electrode forms a transmission line with the beam-pipe wall behind it and has a capacitance and an inductance per unit length, the electrode can resonate with current nodes at each end. Any resonance that has a voltage node at the external connection will resonate with a high Q, because no power can be coupled into the external circuit. The voltage induced by the passing beam thus remains on the electrode and, in some instances, can lead to beam instabilities. For this reason, electrostatic-type pickups of any shape should be avoided if possible. An exception is the button electrode, which is so short that the resonances are at very high frequencies (tens of GHz).

One way to couple to all the resonances in pickups is to make the external signal attachment at one end. In this way all resonances will couple to the external circuit. If, however, the transmission line used for the external circuit has a different characteristic impedance than the electrode itself, there will be a voltage-standing-wave ratio (VSWR) at the connection, and several reflections will be required for the power on the electrode to escape onto the transmission line. This will distort the temporal response of the electrode, which is a concern if the electrode is to be used at high frequencies.

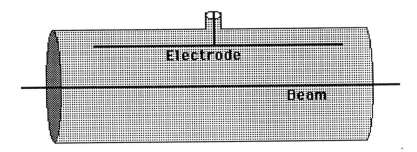

Fig. 9.1. Side view of an electrostatic pickup with the signal connection at the center. Because both ends are open, the beam can excite high-Q resonances in the electrode that have current nodes at the ends and a voltage node at the center.

The standard solution for minimizing the VSWR is to match the characteristic impedance of the electrode and the transmission line. In this case, the power induced on the electrode is transferred to the transmission line without any reflection and, therefore, appears immediately in the external circuit. This design is discussed in the following section.

10. DIRECTIONAL COUPLER PICKUP ELECTRODES

Directional coupler pickup electrodes (sometimes referred to as "stripline" or "microstrip" electrodes) are essentially transmission lines with a well-defined characteristic impedance and with a segment of the center conductor exposed to the beam.

We will examine the signal formation on this type of electrode in both the time and frequency domains. First, we consider the time domain.

Consider an electrode of azimuthal width ϕ, length ℓ, and characteristic impedance Z in a cylindrical beam pipe of radius b, as shown in Fig. 10.1. If the beam current is $I_b(t)$, the wall current intercepted by the electrode is $(\phi/2\pi) \bullet I_b(t)$ for a centered beam. This is the current that flows on the inner surface of the electrode exposed to the beam. When the beam pulse approaches the upstream end of the electrode, this wall current must cross the gap from the beam pipe wall to the electrode. Because the impedance of the gap is half the electrode's characteristic impedance (the inducing current sees two transmission lines in parallel), the voltage induced across the gap is $V(t) = (\phi/2\pi) \bullet (Z/2) \bullet I_b(t)$. This voltage then launches TEM waves in two directions. One signal goes out the upstream port to the electronics. The other wave travels down the outside surface (primarily) of the electrode to the downstream port at a signal velocity $v_s = \beta_s c$ and out the downstream port. Because the impedance of the transmission line is Z, the current flowing in each direction is half the current flowing on the inside surface of the electrode. The beam travels down the beam pipe at a velocity $v_b = \beta_b c$ and induces a similar signal at the downstream gap, with a delay given by $\ell/\beta_b c$ and with opposite polarity to the first one. The net result is that at the upstream port, we see a bipolar-doublet signal of the form

$$V_U(t) = \frac{\phi Z}{4\pi}\left[I_b(t) - I_b\left(t - \frac{\ell}{\beta_b c} - \frac{\ell}{\beta_s c}\right)\right] \; ; \qquad (10.1)$$

while at the downstream port, the bipolar signal is given by

$$V_D(t) = \frac{\phi Z}{4\pi}\left[I_b\left(t - \frac{\ell}{\beta_s c}\right) - I_b\left(t - \frac{\ell}{\beta_b c}\right)\right] \; . \qquad (10.2)$$

If $\beta_b c$ equals the TEM wave velocity $\beta_s c$ on the electrode, there is complete cancellation at the downstream port. Therefore this form of electrode structure can be highly directional. Directivities (the ratio of forward to reverse power) close to 40 dB have been obtained. This is quite useful in monitoring beams in collider-type storage rings that have counter-rotating beams.

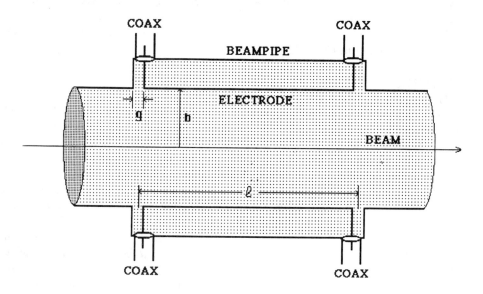

Fig. 10.1. Side view of a directional-coupler beam position electrode. The electrode is a section of an impedance-matched transmission line with the center conductor exposed to the image fields of the passing beam. Signals induced on the electrode exit through the upstream and downstream ports without reflection.

There are several variations of this geometry. In addition to terminating the downstream port in the characteristic impedance, the port may be either shorted to ground or left open. In these cases, the signal traveling down the electrode to the downstream port is reflected, either inverted (shorted electrode) or noninverted (open electrode). In the latter case, a signal induced at the downstream port by the beam is twice the amplitude and the opposite polarity of the reflected signal. In all cases, the output signal at the upstream port is a bipolar doublet with zero net area.

For a Gaussian beam bunch shape, the bipolar doublet is of the form

$$V_U(t) = \frac{\phi Z}{4\pi} \left[\exp\left[\frac{-(t+\tau)^2}{2\sigma^2} \right] - \exp\left[\frac{-(t-\tau)^2}{2\sigma^2} \right] \right] I_b(t) \; , \qquad (10.3)$$

$$\text{where } \tau = \frac{\ell}{2c} \left[\frac{1}{\beta_b} + \frac{1}{\beta_s} \right] \; . \qquad (10.4)$$

This is plotted in Fig. 10.2 for a 10-cm-long electrode with several values of bunch length σ.

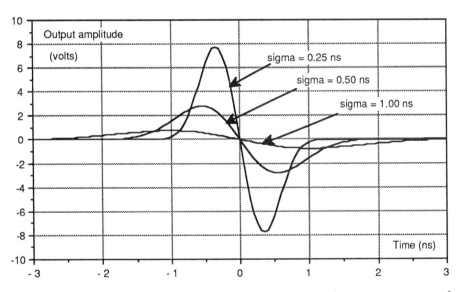

Fig. 10.2. Plot of the expected voltage waveform at the upstream port of a 10-cm-long, 45°-electrode 50 Ω directional-coupler pickup for 10^{10} protons with rms beam bunch lengths of 0.25, 0.5, and 1 ns.

The same analysis can be performed in the frequency domain using Eq. (4.4) for a beam current modulated at frequency $\omega/2\pi$. In this case the rms voltages at the upstream and downstream ports are

$$V_U(\omega) = \frac{\phi Z}{\sqrt{2\pi}} \langle I_b \rangle A(\omega) \sin\left[\frac{\omega \ell}{2c} \left(\frac{1}{\beta_s} + \frac{1}{\beta_b} \right) \right] \qquad (10.5)$$

and

$$V_D(\omega) = \frac{\phi Z}{\sqrt{2}\pi} \langle I_b \rangle A(\omega) \sin\left[\frac{\omega\ell}{2c}\left(\frac{1}{\beta_s} - \frac{1}{\beta_b}\right)\right] \quad . \tag{10.6}$$

Note that there is a length of the electrode for which the argument of the sine function in Eq. (10.5) is $\pi/2$, and the output signal is maximized. This electrode is sometimes referred to as a "quarter wavelength" electrode, even though it is not quite quarter wavelength except when β_b and β_s are both equal to 1. Note also that there are periodic zeros in the response function when the argument is equal to $n\pi$. When this type of pickup is used as a wide-band pickup for looking at signals over a large frequency range, these dips in the response function are easily observed.

At low frequencies, the circular function in the expression for the signal amplitude at the upstream port can be replaced by its argument. The rms output voltage at low frequencies then becomes:

$$V_U(\omega) = \frac{\phi Z}{\sqrt{2}\pi} \langle I_b \rangle A(\omega) \frac{\omega\ell}{2\beta_b c}\left(1 + \frac{\beta_b}{\beta_s}\right) \quad . \tag{10.7}$$

This approximation is valid if the electrode is substantially shorter than a quarter wavelength.

Comparing this approximation to Eq. (8.4) for the button electrode, we see that the two expressions are identical if the signal velocity β_s in Eq. (10.7) is set to infinity and R in Eq. (8.4) is set to $Z/2$. We have used two very different electrode designs in these calculations and two very different calculational methods. When a directional coupler electrode is very short, the output impedance does look like $Z/2$ (because it is terminated in Z at both ends), and we can also ignore the finite signal velocity on the electrode, so this equivalence is expected.

11. OTHER TYPES OF ELECTROMAGNETIC PICKUPS

There are several other types of electromagnetic position pickups that should be mentioned. Small loop couplers[22] (often called B-dot loops, meaning dB/dt) are simply small shorted antennas that couple to the azimuthal magnetic field of the passing beam and can be quite directional.

A second type of position pickup that has been used in the past is the "window frame" pickup electrode.[23] In this design, a ferrite window frame is linked with a conductor so that off-center beams create a difference signal.

A third type is the so-called slot coupler (Faltin pickup[24]) in which the beam fields couple to a nearby center conductor of a transmission line through holes or slots in the ground-plane wall, and induce signals that travel in the beam direction. One shortcoming of this design is that the slots make the transmission line dispersive, and the signal remains in synchronism with the beam for only a narrow frequency range.

A fourth type is a resonant rf cavity excited by a bunched beam in a TM mode that has a null for a centered beam. An off-center beam excites cavity resonances whose amplitudes are proportional to the product of beam intensity × beam displacement and whose phase is dependent on the direction of displacement.[25]

12. HIGH-FREQUENCY EFFECTS

There are two high-frequency effects that the designer should consider when an electrode structure is being designed. These effects are the gap transit time and the Bessel factor.

If the gap g along the direction of the beam between the ground plane and the end of the electrode (see Fig. 10.1) is such that the transit time of the particle across the gap is a significant fraction of the period of the signal being measured, then the resultant signal amplitude is reduced by the transit time factor (TTF)

$$\text{TTF} = \frac{\sin \alpha}{\alpha} \quad ; \alpha = \frac{\omega g}{2\beta_b c} \quad . \tag{12.1}$$

The 3-dB point (TTF $= 0.707$) occurs when α in the above equation is about 1.4 radians (80°).

The Bessel factor arises from the fact that if the particle is not traveling at exactly the speed of light, the EM fields accompanying it are not TEM waves,[22] but have a finite longitudinal extent. This can be seen as follows. Consider a charged particle at rest and centered in a hollow conducting tube. In this case, the field lines connecting the particle to the tube have a finite longitudinal extent (along the axis of the tube), as shown in Fig. 12.1. The longitudinal distribution of charge on the inner wall of the hollow cylinder extends over about $b/\sqrt{2}$ (rms length) where b is the tube radius. If we now transform into a rest frame in which the particle is moving with a velocity $v_b = \beta_b c$, this longitudinal distribution of fields, and the corresponding wall current, moves with it. Note that the wall current actually precedes the particle. For highly relativistic particles, this longitudinal distribution of fields and corresponding wall currents contracts (Lorentz contraction) into a disk, i.e., a TEM wave with no z components. At high frequencies, this finite longitudinal extent causes the signals of slow particles ($\beta_b < 1$) to roll off with the (Bessel) factor BF

$$BF = \frac{1}{I_o(arg)} \quad ; \quad arg = \frac{\omega b}{\beta_b \gamma_b c} \quad , \tag{12.2}$$

where $I_0(arg)$ is the modified Bessel function of order zero. For $arg = 0$, the Bessel factor is 1. The 3-dB point occurs for $arg = 1.22$. Good design practice generally limits arg to less than about 1.

As an example, consider a 1-MeV ($\beta_b = 0.046$) proton beam in a pickup with a 1 cm half-aperture. The 3-dB point occurs at about 270 MHz.

13. SIGNAL-TO-NOISE AND RESOLUTION

The available thermal noise power in a bandwidth B at temperature T is given by $P_N = kTB$ where k is Boltzmann's constant, T is the temperature in kelvin, and B is the bandwidth in hertz. This is about -114 dBm per MHz

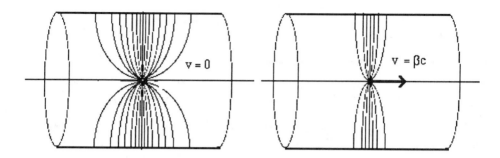

Fig. 12.1. The longitudinal field distribution of a static and moving charge ($\beta = 0.9$) in a grounded conducting cylinder. The longitudinal distribution contracts to a flat disk for highly relativistic particles. For slow particles, the field lines extend about $\pm b/\sqrt{2}$ longitudinally both in front of and behind the particle.

at room temperature. If we include the noise figure of the electronics, the noise power is often many decibels higher.

Resolution is sometimes limited by the noise power. We can use the relations developed in Eqs. (5.5) or (5.6) to estimate the resolution limit caused by thermal noise. The resolution limit is

$$\frac{\delta x}{b} = \frac{1}{4\sqrt{2}} \frac{\phi}{\sin(\phi/2)} \sqrt{\frac{P_N}{P_S}} \approx \frac{1}{2\sqrt{2}} \sqrt{\frac{P_N}{P_S}} , \tag{13.1}$$

where P_S is the signal power on a single electrode. Hence, a 40-dB signal-to-noise ratio (100:1 amplitude ratio) limits the resolution to about 0.35% of the half aperture.

As an example, consider a 1-mA beam in a quarter wavelength 45° (ϕ=0.79 radian), 50-Ω pickup. The signal power P_S per electrode at the upstream port of a directional coupler is given by

$$P_S = 2\left(\frac{\phi}{2\pi}\right)^2 Z \langle I_b \rangle^2 A^2(\omega) \sin^2\left[\frac{\omega\ell}{2c}\left(\frac{1}{\beta_b} + \frac{1}{\beta_s}\right)\right] . \tag{13.2}$$

Using the above parameters, $P_S = 1.56 \ \mu W$, which corresponds to -28 dBm. With a 10-MHz bandwidth and an electronic noise figure of 6 dB, the noise power P_N is -98 dBm. Thus the signal-to-noise ratio is about 70 dB, and the resolution limit is about 3 μm for a 50-mm-diam aperture.

Note that the signal power scales linearly with the characteristic impedance of the pickup, as is expected for current sources. Thus if signal-to-noise is a problem, raising the pickup impedance is a solution. Pickups with impedances exceeding 100 Ω have been used successfully. One way to accomplish this over a narrow band is to use a quarter-wave (transmission line) transformer at the pickup to match the pickup to the cable. Resonant beam position monitors with effective electrode input impedances of approximately 5000 Ω have been made for measuring the position of very small beam currents.[26]

The RFI from accelerator rf systems can have serious effects on the accuracy and resolution if the shielding is not adequate. If the beam-bunching frequency is a subharmonic of the rf frequency, then operating the pickup at the lower frequency often eliminates the interference. EMI (electromagnetic interference) from pulsed and SCR power supplies is often a problem, and great care must be taken to eliminate ground loops that can pick up noise.

Shot noise is not a noise in the case of particle beams, but a signal. It relates specifically to fluctuations in the instantaneous beam current caused by the granularity of the individual particle charges. Because the fluctuations in beam current create a carrier signal, they allow detecting the beam position and therefore are sometimes called Schottky currents. The rms Schottky current for a coasting beam of average current $\langle I_b \rangle$ and bandwidth B is (for particles with charge $\pm e$)

$$I_{\text{shot}} = \sqrt{2e\langle I_b \rangle B} \ .$$
(13.3)

A specific example is that the Schottky current for a 1-A proton beam with a 1-MHz bandwidth is 0.6 μA.

14. SIGNAL-PROCESSING METHODS

There are three general methods for deriving a normalized position signal from the raw pickup electrode signals. They are difference-over-sum (Δ/Σ) processing, amplitude-to-phase conversion (AM/PM) processing, and log-ratio processing. Each of these have certain advantages over the others. Specific advantages to look for are simplicity and cost, dynamic range (of beam signal intensity), linearity (of response vs displacement), and bandwidth (either acquisition or real-time).

Figure 14.1 shows the expected amplitude response using these three signal-processing methods for a 45° electrode width in a circular beam pipe. It is apparent that of the three processing methods, the log-ratio technique yields the response most linear in beam position.

15. DIFFERENCE-OVER-SUM PROCESSING

There are many variations of the difference-over-sum (Δ/Σ) method. We define shorthand notation as

$$\frac{\Delta}{\Sigma} = \frac{I_R - I_L}{I_R + I_L}$$
(15.1)

This ratio is "normalized," because it is independent of the beam current.

The simplest approach to obtain Δ/Σ would be to detect the rf signals using diode detectors, homodyne detectors,[27] or demodulator chips,[28] and then to generate analog signals proportional to the rf envelope amplitudes. These signals

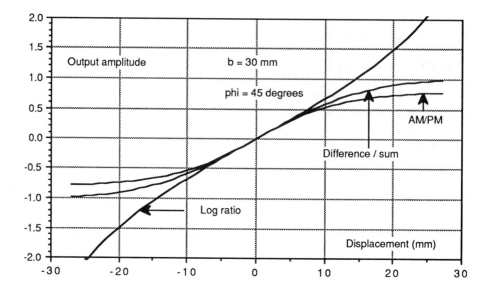

Fig.   14.1.   Plot of the response of difference-over-sum, amplitude-to-phase conversion, and log-ratio processing methods to a displaced beam in a 60-mm-diam aperture with 45° directional-coupler electrodes. The log-ratio response is the most linear. All three curves have been normalized to the same slope at the center.

can then be digitized and then processed, or processed in analog circuits to achieve a Δ/Σ signal before digitization. Another common approach is to use 180° hybrid junctions to generate the Δ/Σ signal before rf detection. This latter method requires that the signals, and hence the signal cables, be properly phase-matched at the hybrid junction.

If the digitization is performed before the Δ/Σ function, the granularity of the ADC limits the dynamic range of the system. If the beam current varies over a 100 to 1 dynamic range (40 dB), the resolution with a 12-bit ADC is limited to about 2% of the half-aperture of the pickup at the low end. If the Δ/Σ function is performed in analog circuits, the analog-division process is slow and has a narrow dynamic range. An alternate process uses an automatic-gain-control (AGC) circuit as a normalizer.[29] This particular circuit also uses switching between electrodes at 40 kHz to eliminate possible gain differences between rf amplifier channels, which also limits the bandwidth.

One advantage to the Δ/Σ process is that it can also be carried out in the time domain by using a peak detector to capture the peak voltage in the bipolar signal from the individual electrodes.[30] A disadvantage of peak detection is that the peak voltage is very sensitive to the pulse shape, as is seen in Fig. 10.2. It is also very sensitive to cable attenuation and dispersion in long cables.[31] Of all the processing methods, Δ/Σ is perhaps the easiest to implement, and therefore is by far the most popular method.

16. AMPLITUDE-TO-PHASE CONVERSION

In the amplitude-modulation-to-phase-modulation (AM/PM) method,[32] the two in-phase rf signals from the two pickup electrodes are split and re-combined in quadrature (i.e., with 90° relative phase shifts at the processing frequency) to convert the two signals of different amplitudes into two equal-amplitude signals whose phase difference is related to the amplitude difference of the two incoming signals. Figure 16.1 shows a block diagram of the basic conversion process, along with phasors indicating the signal processing. The transfer function is given by

$$\Delta\theta = 2 \tan^{-1}\left(\frac{R}{L}\right) - \frac{\pi}{2} \qquad (16.1)$$

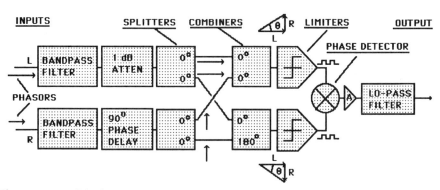

Fig. 16.1. Block diagram of the basic processing system for amplitude-to-phase conversion. After the two in-phase signals are filtered, they are split and recombined in quadrature using power splitters and combiners. At this point, the amplitude difference of the two input signals has been converted to a phase difference (6.6° per dB). The limiters remove the amplitude variation while preserving the phase information. The phase detector produces an analog signal proportional to the phase difference.

The conversion gain for these circuits is typically 6.6° of phase shift for each decibel of amplitude difference in the incoming signals. The quadrature phase shift can be accomplished by using either quarter-wavelength transmission lines (least expensive, but limited to a single frequency) or quadrature hybrids. The two signals are then clipped to a constant amplitude in a hard-limiter circuit that preserves the phase information within a fraction of a degree but removes all amplitude dependence, and then the phase difference is measured in a phase-detector circuit. The most common hard limiter presently in use is the 9685-type comparator, and the phase detectors are usually double-balanced mixers or exclusive ORs.

The analog voltage output of the AM/PM processor is related to the amplitudes R and L of the two input signals by the relation

$$V_{\text{out}} = V_o \left[\tan^{-1}\left(\frac{R}{L}\right) - \frac{\pi}{4} \right] = V_o \tan^{-1}\left(\frac{R-L}{R+L}\right) . \qquad (16.2)$$

Note the relationship to the Δ/Σ processing algorithm.

Hard limiters are usually the components that most restrict the system performance. Their dynamic range varies from 40 to over 60 dB depending on the operating frequency. They have been used at frequencies up to 100 MHz but perform best below about 20 MHz. To use hard limiters when the beam-current modulation is over 100 MHz, the rf signal is often heterodyned down to a lower frequency, typically 10 or 20 MHz. The real-time bandwidth is about 10% of the processing frequency and can be as high as 10 MHz.

AM/PM circuits using down conversion from 200 to 10 MHz and 425 to 20 MHz have been built at Los Alamos.[33] Fermilab has built approximately five hundred 53-MHz units and has built down-conversion units for both the 200-MHz proton linac and for the rapid-cycling Booster synchrotron.[34] This latter application is quite interesting in that the local oscillator used for the down-conversion tracks the Booster rf frequency from 30 to 53 MHz with a constant frequency offset, so that the IF frequency output to the AM/PM circuit is constant.

It is essential to keep the relative phase of the two signals entering the AM/PM processor within about ±5°, otherwise Eq. (16.1) must be modified. Phase matching of the cables from the pickups to the electronics is therefore required.

The AM/PM circuits are specifically for frequency-domain signal processing and therefore function well for multibunch beam pulses. AM/PM processing has also been used for the measurement of the position of single, isolated beam bunches. In this application, a narrow-band filter is placed upstream of the AM/PM processor. The passage of a single bunch by the pickup electrode shock-excites the bandpass filter to ring for perhaps 10 cycles, which is sufficient time to complete the phase difference measurement.[21, 32] These filters must be very carefully matched, with center frequencies equal within about ±0.1%. Both lumped-component circuits and shorted coaxial transmission lines have been used in this application.

Real-time bandwidth is important if the signal is to be used in real-time control of the beam position. In this case, the processed beam position signal is fed back to a beam deflector electrode located upstream of the pickup, and the system is run in closed-loop fashion to reduce the transverse beam jitter. Because AM/PM system provides a normalized position signal, the position signal is independent of beam intensity. The jitter-control function can be accomplished more simply by using an unnormalized difference signal if the beam intensity variation is minimal. In this case, the closed-loop gain is proportional to the beam intensity.

Of the three basic types of position processors reviewed here, the AM/PM circuit is the most expensive and difficult to implement. However, the obtainable large dynamic range and high real-time bandwidth are very desirable features. Thus far, their application is limited to the Linac, Booster, Main Ring, and Tevatron[35] at Fermi National Accelerator Laboratory (FNAL); the Proton Storage Ring at Los Alamos National Laboratory;[36] and the LEP ring[21] at CERN. The HERA-P ring at DESY also is planning to use a similar circuit. The SLAC SLC arcs[37] use Δ/Σ, however.

If ultimate resolution is the main objective, it is important to consider the effective noise figures obtainable with the various circiuts. The effective noise figure for the electronics at the input of the hard limiter in AM/PM processing is in the 10- to 20-dB range and, in addition, there is noise downstream of the hard limiter that produces a resolution minimum independent of beam current.

17. LOG-RATIO PROCESSING

In the log-ratio method, the two signals are put into a discrete or hybrid circuit that produces an output signal proportional to the logarithm of the ratio of the two input signals. The log function is usually obtained by using an ideal forward-biased diode junction in the feedback circuit. There are several limitations to this method. First, commercial circuits are limited to a few hundred kilohertz of bandwidth because of the high dynamic resistance of the "diodes" (actually the base-emitter junctions of matched transistors) at low currents in combination with the distributed capacitance. A circuit designed for high-bandwidth operation possibly could achieve a bandwidth approaching 10 MHz. Second, the log-ratio circuit inputs are usually unipolar, so the signals must be the detected-envelope signals rather than the raw rf signals. Third, the output signal amplitude is proportional to the absolute temperature of the diode junction and should be temperature-compensated. The Analog Devices[38] AD538 is a commercially available log-ratio chip with temperature compensation. The Analog Devices AD540 chip uses controlled gain compression to emulate the log process and has bandwidths approaching 100 MHz.

The output of the log-ratio circuit is proportional to

$$V_{\text{out}} = V_o\, Ln\left(\frac{R}{L}\right) = 2V_o\, \tanh^{-1}\left(\frac{R-L}{R+L}\right) \qquad (17.1)$$

Note the similarity to both the Δ/Σ and the AM/PM algorithm.

18. INTENSITY MEASUREMENT

Although the pickups and circuits discussed here are primarily designed for position measurement, they can be used to measure beam intensity as well. An example of this is the beam position system in the FNAL Tevatron, which has about 220 beam position monitors evenly distributed around the ring (mostly at 4 kelvin because the machine is superconducting) and only one beam current monitor. During the commissioning process, the ability to measure the beam intensity at every beam position monitor (using the sum of the electrode signal amplitudes) with a few percent accuracy was absolutely necessary to obtain the first full revolution of beam. Because the output power of the summed signal is slightly sensitive to both the beam bunch length and the beam position, the accuracy is limited.

19. ALIGNMENT, CALIBRATION, AND ON-LINE TESTING

The initial alignment and calibration can be separated into three main components: the pickup electrode assembly, the cabling, and the electronics.

After the pickup construction and assembly is complete, the electrode assembly is normally first "mapped" with an rf-excited taut wire stretched through the pickup at specific points to determine the electrical response for various displacement values of x and y, to determine both the electrical center and the displacement sensitivities, as well as the nonlinearities.[19] The pickup is then installed and aligned mechanically with the other beamline components. An optional alternative is to place the taut wire along the magnetic axis of the beamline magnetic optics after the pickup is mechanically mounted, excite the wire with an rf signal, then measure the electrical offset (decibels of output signal unbalance) relative to the magnetic, rather than the mechanical, axis of the beam optics.[39] The electrical offset and the displacement sensitivity are then put into the computer database for correction to the beam position data as they are read out. The nonlinearities, as shown in Fig. 14.1, can be corrected in a look-up table at this time.

In the taut-wire measurement, the signal on the wire is by definition a TEM (principal) wave with no longitudinal electric or magnetic component, travels at the velocity of light, and emulates highly relativistic particles. The signals associated with slow particles, as discussed in Sec. 12 and illustrated in Fig. 12.1, are very difficult to generate on wires. As was shown in Secs. 6 through 10, the pickup response depends on the particle velocity, and this must be taken into account when calculating the pickup response.

The coax cables connecting the pickup electrode to the electronics must be checked for attenuation. If the signal processing is dependent on the relative phase delay in the cable pairs, their electrical length must be equalized as well.

The electronics can be separately calibrated for any electrical offset in output reading and in the voltage output vs decibel difference between the two inputs at a variety of fixed signal input power levels. Because the electronics is an active circuit, it can be expected to have nonlinearities. Therefore the electronics needs to be "aligned" before installation. The alignment process can be quite cumbersome if performed manually, so it is often performed using a computer-automated system. A common alignment circuit in use at both FNAL and LANL is an rf signal generator that uses electronic attenuators under computer control to step through a variety of input unbalances (-10 dB to $+10$ dB in 1-dB steps, and from -60 dBm to 0 dBm in total power output in 1-dB steps (a total of 1260 settings) in about 100 ms.[33] The electronic unit also provides a z axis and a horizontal axis signal (linear in total power dBm) to an x-y oscilloscope. The analog output of the position-processing electronics provides the y-axis signal. The oscilloscope display is thus a real-time raster display with a 10-Hz refresh rate.

The overall calibration of assembled systems is of the order of 0.5% to 1% of the full aperture for center position accuracy and about $\pm 2\%$ on the displacement sensitivity S_x. Resolution can approach the signal-to-noise limit in Eq. (13.1).

Equally important is providing some way in which to monitor the performance and calibration of a system while it is in use. A method used at FNAL provides special test inputs into the rf input ports of the electronics that could perform these tests. A 10-mA dc current was put through a high-frequency inductor and onto the upstream side of a dc isolator in the rf circuitry. Because the pickup electrodes were back-terminated directional couplers, a measurement of the dc voltage at the point of current input determined the back-termination value within a few ohms and that continuity existed through the signal cable and the directional coupler (at 4 kelvin in the Tevatron) to the back terminations. The same test ports allowed injection of rf signals through a capacitive coupling into the signal-processing electronics with several decibel ratios to check proper operation of the electronics. All these tests, as well as the digitizers for collecting the measurement data, were under local microprocessor control, and the measured data was stored in RAM for subsequent readback by the host computer.

The degree of automation required depends on the size of the position measurement system, and the remoteness of the system components from the central computer. Because beam position measuring systems normally do not have redundancy, it is important to be aware of any channels that are not operating properly.

20. GENERAL HARDWARE SYSTEM ARCHITECTURE

More often than not, accelerator control systems, of which the beam diagnostics systems are an integral part, utilize distributed intelligence to control, read back, and perform a set of on-line tests on the systems that interface to the accelerator.

There are several advantages to using distributed intelligence. First, if a particular diagnostic subsystem performs a critical protection function on an accelerator component, the subsystem must remain operational even when the network to the host computer fails. Second, the microprocessor contains in its database the programs for performing the data collection tasks and the on-line calibration tasks, which can be initiated by the host and performed without host intervention.

In small, local systems, such complexity is not needed. An issue that always seems to arise in the design of small systems is whether the simultaneous measurement of beam position at many position pickups is required. It is certainly much less expensive to install multiplexers and sequentially process the signals from different pickups. However, if the accelerator has a very low duty cycle and a low repetition rate, data collection can be very slow with a multiplexed system. Also, if it is important to record betatron oscillations around the closed orbit in a synchrotron for "tune" or betatron phase advance information, a nonmultiplexed system is much preferable.

In terms of interconnections to the host, the following scheme was found to be quite efficient at the Tevatron.[40] A high-speed communication network (in this case serial CAMAC) was used as a point-to-point bidirectional link for downloading control parameters and reading back data asynchronously (not in real time but under the control of the host). A second single coaxial cable carried a 10-MHz bipolar clock signal with a set of superimposed diphase-encoded timing signals simultaneously to all microprocessors to synchronize them and to initiate sample-and-hold gating and digitization of beam position information or on-line system tests. The results were then stored in local memory until the host requested it. This minimized the traffic on the computer network.

21. SOFTWARE AND DISPLAYS

The effort put into developing software and displays always seems to be underestimated. Creative programmers never seem to have sufficient time to implement their plans. There is however, a minimum amount of software that is truly necessary. It usually is very unique to the specific application and cannot be specified adequately without a detailed knowledge of the system requirements.

Nevertheless, it is possible to say a few words about certain common aspects of software and displays. Automatic beamline control by closing the control loop from the position pickup to beam steering devices through the host computer is becoming more necessary as accelerators and beamlines become more complex. It simply is not possible to steer a beam through a long, complex beamline, or to minimize the orbit distortions in a large synchrotron, without the support of software algorithms.[41,42] In some beamlines, it may be more accurate to measure the magnetic-lattice transfer matrices on-line than to compute them from beamline lattice designs.

In another area, the need for archiving measurement data for later recall in order to make comparisons is very important in locating malfunctioning accelerator components. For example, "cusps" in the orbit distortions that pinpoint the orbit perturbation are determined by subtracting archived data from new data for the same accelerator operating conditions. Another application is making specific measurements that require changing the accelerator operating conditions; (e.g., making measurements of the dispersion in the lattice[43]). A third reason for archiving on a very short-term basis is to store a series of beam position measurements in FIFO memories thereby making available instant replays of accelerator performance just before unexpected equipment failures and subsequent beam loss. This of course represents a tremendous amount of data and certainly cannot be handled over a network. For this reason, such a capability must be built into the distributed intelligence, local to the beam position electronics.

Having the right software and displays available at turn-on is a tremendous asset in commissioning a facility. The software is usually one of the last components to be "installed," and as often as not, it is written by a programmer who is unfamiliar with the accelerator and therefore does not write programs that provide the displays and other information in the most useful format for presentation. Furthermore, as distributed intelligence is becoming more common, the software "tree" is becoming multitiered in that specific control and data-processing functions are being separated into various locations, often with different operating systems, programming languages, and even programmers.

22. CONCLUSIONS

In designing a beam position measurement system, it is important to understand fully the characteristics of beam position diagnostics devices and the accelerator facility in which they are to be installed. It is possible to build a technically excellent piece of hardware that relies on the presence of beam modulation signals that are not always present, or does not take into account all the possible operating conditions of the accelerator, and therefore does not always provide the required measurements. In short, the designer must be knowledgeable in all aspects of the accelerator operation, the beam-coupling mechanism to the pickup electrodes, electronic instrumentation design, and in the applications software.

23. REFERENCES

1. J. Borer and R. Jung, "Diagnostics," CERN Publication 84-15, 385 (1984).

2. G. Lambertson, "Dynamic Devices; Pickups and Kickers," AIP Conf. Proc. **153**, M. Month, Ed., 1413 (1987).

3. R. Littauer, "Beam Instrumentation," AIP Conf. Proc., **105**, M. Month, Ed., 869 (1983).

4. R. H. Siemann, "Bunched Beam Diagnostics," AIP Conf. Proc. **184**, M. Month, Ed., 430 (1989).

5. J. L. Pellegrin, "Review of Accelerator Instrumentation," Proc. XI*th* International Conf. on High-Energy Accelerators, CERN, 459 (1980).

6. "Frontiers of Particle Beams; Observation, Diagnosis, and Correction," Proc. of the Joint U.S. CERN School on Particle Accelerators, M. Month and S. Turner, Eds., Capri, 1988, Springer Verlag Lecture Notes in Physics **343** (1989).

7. C. D. Moore et. al.; "Single Bunch Intensity Monitoring System Using an Improved Wall Current Monitor," Proc. 1989 Particle Accelerator Conf., IEEE Catalog No. 89CH2669-0, Chicago, IL, 1513 (1989). See also R. Webber, these proceedings.

8. R. Jung, "Beam Intercepting Monitors," Ref. 6, page 403.

9. R. J. Nawrocky et al., "Automatic Control of Position and Direction of X-Ray Beams at NSLS," Proc. 1989 Particle Accelerator Conf., IEEE Catalog No. 89CH2669-0, Chicago, IL, 1856 (1989).

10. D. D. Chamberlin, "Imagescope to Photodiode Beam Profile Imaging System," IEEE Trans. Nucl. Sci. **30**, 2201 (1983).

11. F. Hornstra, "Residual Gas Ionization Profile Monitor for HERA," Proc. European Particle Accelerator Conference, Rome, 1160 (1988).

12. C. Field, "Problems of Measuring Micron Size Beams," Proc. 1989 Particle Accelerator Conf., IEEE Catalog No. 89CH2669-0, Chicago, IL, 60 (1989).

13. R. Ericson, "Monitoring in Future e^+e^- Colliders," Ref. 6, page 482.

14. W. B. Cottingame et al., "Noninterceptive Monitoring of Longitudinal Parameters in H^- Beams," IEEE Trans. Nucl. Sci. **32**, 1871 (1985).

15. M. Placiti and R. Rossmanith, "e^+e^- Polarimetry at LEP," Nucl. Instr. and Meth. **A274**, 79 (1989).

16. W. Barry, R. Rossmanith, and M. Wise, "A Simple Beam Position Monitor System for CEBAF," 1988 Linear Accelerator Conference Proceedings, CEBAF-Report-89-001, 649 (1989). See also P. Adderley et al., "A Beam Monitor for Low Intensity Beams," Proc. 1989 Particle Accelerator Conf., IEEE Catalog No. 89CH2669-0, Chicago, IL, 1602 (1989).

17. R. E. Shafer, "Characteristics of Directional Coupler Beam Position Monitors," IEEE Trans. Nucl. Sci. **32**, page 1933 (1985).

18. K. Satoh, "Beam Position Monitor using Wall Currents," Rev. Sci. Instr. **4**, 450 (1979).

19. W. Schutte, "Results of Measurements on the HERA Proton Beam Monitors," Proc. 1989 Particle Accelerator Conf., IEEE Catalog No. 89CH2669-0, Chicago, IL, 1471 (1989).

20. T. Linnecar, "High Frequency Longitudinal and Transverse Pickups used in the SPS," CERN SPS/ARF/78-17 (1978).

21. J. Borer et al., "LEP Beam Orbit System," Proc. 1987 Particle Accelerator Conf., IEEE Catalog No. 87CH2387-9, 778 (1988).

22. J. H. Cuperas, "Monitoring of Beams at High Frequencies," Nucl. Instr. and Meth. **145**, 219 (1977).

23. J. Claus, "Magnetic Beam Position Monitor," Proc. IEEE Particle Accelerator Conference, NS-20, **3**, 590 (1973).

24. L. Faltin, "Slot-Type Pickups," Nucl. Instr. and Meth. **148**, 449 (1978).

25. J. McKeown, "Beam Position Monitor using a Single Cavity," IEEE Trans. Nucl. Sci. **26**, 3423 (1979). See also ibid. **28**, 2328 (1981).

26. Q. Kerns et al., "Tuned Detector for Fermilab Switchyard," Proc. 1987 Particle Accelerator Conf., IEEE Catalog No. 87CH2387-9, 661 (1988).

27. R. Bossart et al., "Synchronous Receivers for Beam Position Measurement," IEEE Trans. Nucl. Sci. **32**, 1899 (1985).

28. J. Hingston, J. Johnston, and I. Ko, "Advanced Light Source (ALS) Beam Position Monitor," Proc. 1989 Particle Accelerator Conf., IEEE Catalog No. 89CH2669-0, Chicago, IL, 1507 (1989).

29. R. Biscardi, J. W. Bittner, "Switched Detector for Beam Position Monitor," Proc. 1989 Particle Accelerator Conf., IEEE Catalog No. 89CH2669-0, Chicago, IL, 1516 (1989).

30. R. E. Meller, D. Sagan, and C. R. Dunnam, "Beam Position Monitors for the CESR Linac," Proc. 1989 Particle Accelerator Conf., IEEE Catalog No. 89CH2669-0, Chicago, IL, 1468 (1989).

31. The skin effect causes both the signal attenuation and the signal velocity to have a square-root-of-frequency dependence. The attenuation in nepers is equal to the phase shift in radians (i.e., 1 dB of attenuation corresponds to 6.6° of phase shift).

32. S. P. Jachim, R. C. Webber, and R. E. Shafer, "RF Beam Position Measurement for Fermilab Tevatron," IEEE Trans. Nucl. Sci. **28**, 2323 (1981).

33. F. D. Wells and S. P. Jachim, "A Technique for Improving the Accuracy and Dynamic Range of Beam Position Monitors," Proc. 1989 Particle Accelerator Conf., IEEE Catalog No. 89CH2669-0, Chicago, IL, 1595 (1989).

34. R. Webber et al., "Beam Position Monitoring for the Fermilab Booster," Proc. 1987 Particle Accelerator Conf., IEEE Catalog No. 87CH2387-9, 541 (1988).

35. R. E. Shafer, R. E. Gerig, A. E. Baumbaugh, and C. R. Wegner, "Tevatron Beam Position and Beam Loss Monitor Systems," Proc. XIIth International Conf. on High-Energy Accelerators, Fermi National Laboratory, 609 (1983).

36. E. F. Higgins and F. D. Wells, "A Beam Position Monitor System for the Proton Storage Ring at LAMPF," IEEE Trans. Nucl. Sci. **28**, 2308 (1981).

37. J. L. Pellegrin and M. Ross, "Beam Position for SLAC SLC Arcs," Proc. 1987 Particle Accelerator Conf., IEEE Catalog No. 87CH2387-9, 673 (1988).

38. Analog Devices Inc., Two Technology Way, Norwood, MA 02062.

39. Q. Kerns et al., "RF Precision Alignment of Beam Position Monitors in Tevatron," IEEE Trans. Nucl. Sci. **30**, 2250 (1983).

40. R. J. Ducar, J. R. Lackey, and S. R. Tauser, "FNAL Synchronization Control for Collider Operation," Proc. 1987 Particle Accelerator Conf., IEEE Catalog No. 87CH2387-9, 1937 (1988).

41. J.-P. Koutchouk, "Trajectory and Closed Orbit Correction," Reference 6, page 46 (1989).

42. R. Raja, A. Russell, and C. Ankenbrandt, "The Tevatron Orbit Program," Nucl. Instr. and Meth. **A242**, 15 (1985).

43. R. E. Shafer, "Beam Diagnostics at the Tevatron," IEEE Trans. Nucl. Sci. **32**, 1862 (1985).

BEAM PROFILE MEASUREMENTS*

John Galayda

National Synchrotron Light Source
Brookhaven National Laboratory
Upton, New York 11973

Abstract

The physics and practical performance limits of various types of profile monitors are described. These devices essentially all detect the products of electromagnetic interactions of the particle beam with solid, gaseous or even liquid targets. These mechanisms can be more or less destructive to the beam. Low emittance and high luminosity beams have prompted improvements in resolution to 30 microns and below.

Introduction

The purpose of beam diagnostic equipment and techniques is to measure various moments of the charge distribution in the beam. Current monitors measure the integral over space and time of the charge distribution. Electromagnetic pickup devices like striplines measure the location of the center of charge of the beam, or the first moments of the transverse charge distribution, $<x>$ and $<y>$. They also give information on the longitudinal extent and shape of the bunch. Beam profile monitors are used primarily to measure the transverse shape of the beam, e.g. the moments $<x^2>$, $<y^2>$, and $<xy>$. Some applications of profile monitoring techniques are not really suitable for a scholarly tome; for example, fluorescent flags may be used to determine if there is any beam anywhere in the general vicinity of the beampipe, in those rare instances (usually very late at night) when the uncertainties as to the proper setting of the beam transport optics grow past the 10% level. No further text will be devoted to this extreme case; knowledge of the basic properties of profile monitors, combined with sufficient desperation, will lead the reader to acceptable techniques for such situations. The connection of beam profile

---

* This work was performed under the auspices of the U.S. Department of Energy.

information to transverse emittance is covered thoroughly in O.R. Sander's lecture. Here we will address the basic mechanisms that produce a signal from a passing beam; the electromagnetic interaction of a beam with matter. We will then consider various destructive and non-destructive measurements of beam profile in transport lines and synchrotrons. These may be broadly categorized in terms of what sort of matter is placed in the beam path (a solid material or perhaps a gas) and what kind of signal is produced (charges from ionization in the detector material or light from fluorescence of the material). Particles of sufficiently high energy (in units of rest mass) produce synchrotron radiation as they pass through a bending magnet, providing a beam profile signal ready for imaging. This phenomenon has become increasingly important as the number of electron accelerators increases. Proton synchrotrons are now attaining energies at which synchrotron radiation becomes significant.

Electromagnetic Interaction of the Beam with Matter

The full description of the interaction of a particle beam with matter in its path must cover many phenomena; ionization of loosely and tightly bound electrons, delta (δ) rays, bremsstrahlung, pair production, Cerenkov radiation, and nuclear (strong) interactions. However, the signals we see in our beam diagnostic equipment originate (for the most part) from small-angle Coulomb scattering of the beam particles with the nuclei of the intervening matter. For this reason we will dwell a bit on the formulas describing Coulomb scattering, following Ritson's book.[1] Complete derivations may be found in several sources;[2] here we will discuss the phenomena and the formulas that describe them in an effort to give the reader enough information to estimate signal strengths and to illustrate some important features like the "minimum ionizing" regime and the concept of radiation length.

The classical cross section for scattering of a charged particle from a very massive point charge like a nucleus is

$$\sigma\,(\theta) = \frac{z^2\,Z^2\,e^4}{4\,p^2\,\beta^2\,c^2}\,\frac{1}{\sin^4\,(\theta/2)}\,d\Omega \tag{1}$$

Where z is the charge of the beam particle; Z is the charge of the target nucleus; θ is the scattering angle; β is the ratio of particle velocity to the velocity of light in a vacuum; p is the momentum of the particle beam; and $d\Omega$ is the increment of solid angle area of a ring-shaped strip on a sphere of unit radius through which pass particles that have been scattered through angles close to θ:

$$d\Omega = 2\pi \sin (\theta) \, d\theta$$

One may think of this cross section $\sigma(\theta) d(\Omega)$, Fig 1, as an expression of the area of a ring-shaped region around a target nucleus. Beam particles passing through this region will be scattered through an angle θ. The above formula correctly accounts for the effect of relativistic quantum mechanics if the incident particle has spin zero. For incident spin 1/2 particles like electrons, it is necessary to average this cross section over the spins of the incoming particles and sum over the spins of the outgoing particles. To account for this one must multiply the above formula by a factor

$$1 - \beta^2 \sin^2 (\theta/2)$$

Obviously this term has little effect for small-angle scattering. The cross section itself is especially large for small values of θ; so most of the charges and photons that we will detect with our profile monitors will be produced in small-angle scattering events. We may therefore use small angle approximations in computing scattering rates. Given the cross section formula we may compute the probability that a particle passing through a thickness x of target material is scattered through an angle in the range θ to $\theta + d\theta$. This probability is just the ratio of the sum of cross sections $\sigma(\theta)$ for all the N target nuclei, divided by the area of the target presented to the beam. This area is the volume of the target divided by its thickness.

$$dP (\theta) = \frac{Nx}{Vol} \, \sigma(\theta) \, 2\pi \sin \theta \, d\theta \qquad (2)$$

For small angle scattering we may approximate $\sin(\theta)$ by θ. Given the above probability function we can now compute the mean square value of the scattering angle θ:

$$< \theta^2 > = \int \theta^2 P(\theta) \, d\theta \qquad (3)$$

$$< \theta^2 > = 8\pi < \frac{NZ^2}{Vol} > z^2 \frac{e^4 x}{(p\beta c)^2} \int \frac{d\theta}{\theta} \qquad (4)$$

$$\int_{\theta}^{\theta_{max}} (d\theta/\theta) = \ln(\theta_{max}/\theta_{min}) \qquad (5)$$

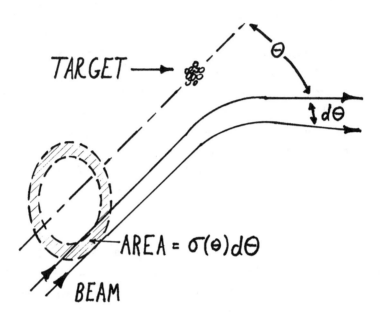

Figure 1. Physical significance of scattering cross section. Particles passing through the ring-shaped region with area $\sigma(\theta) \, d\Omega$ are deflected by an angle between θ and $\theta + d\theta$.

We see the logarithmic angle dependence typical of the Coulomb scattering processes that create the signals used to detect particle beams. The mean square scattering angle depends on the limits of this integral. To go beyond this point we must make some estimate of the values of the minimum and maximum scattering angle, based on some simplified model of the atoms in the target. The logarithm is a very slowly varying function of its argument, so the number we are trying to calculate does not depend critically on our choice of limits. However we cannot extend the integral to zero scattering angle, because the logarithm blows up to infinity. The reason we are faced with this infinity is because some physics has been left out of the problem so far. Our calculation applies to the scattering of two point charges. In reality, however, the scattering target nucleus is surrounded by a cloud of bound electrons that shield the nuclear Coulomb force. A particle that does not penetrate this cloud will not be scattered by the nucleus.

We will not try to compute the force exerted by the electron cloud; we will only estimate the radius of the cloud to set the minimum scattering angle, the lower limit of the integral. A sensible choice of the lower limit comes from the Thomas-Fermi model of the atom, which treats the motion of the electrons as if they were a gas confined by the Coulomb potential[3] :

$$r_{max} = A^{-1/3} .529 \text{ Angstroms} \qquad (6)$$

The nucleus is not a point charge, so if the incident particle comes within a distance r_{min} of the nucleus it will not feel the full Coulomb force; it will be passing within the nuclear charge distribution. The charge radius of a proton or neutron is about 10^{-15} meters, so it should not be surprising that the radius of a nucleus consisting of A protons and neutrons is about

$$r_{min} = A^{1/3} 1.2 \times 10^{-15} \text{ meters} \qquad (7)$$

The empirical constant 1.2 is applicable for incident electrons; 1.4 is more accurate for a proton beam.[4] Small scattering angles are related to these radii so that

$$\theta_{max}/\theta_{min} = r_{min}/r_{max}$$

We can make the approximation $A \sim 2Z$ so that

$$< \theta^2 > = 16\pi < \frac{NZ^2}{Vol} > \frac{z^2 e^4 x}{(p\beta c)^2} \ln (183Z^{-1/3}) \qquad (8)$$

The rms scattering angle is commonly written in terms of the radiation length L_r of a substance. The radiation length is the thickness of material required to degrade the energy of an electron beam by a factor $1/e$. It provides a convenient way to parametrize electromagnetic processes such as pair production and bremsstrahlung. The radiation length of various substances may be found in the Particle Data

Tables.[5] Here the rms scattering angle is expressed as

$$\langle\theta^2\rangle^{\frac{1}{2}} = \frac{14.1 \text{ MeV/c}}{p\beta} \; z(x/L_r)^{\frac{1}{2}} \; \{1 + \frac{1}{g} \log_{10}(x/L_r)\} \text{ (radians)} \qquad (9)$$

By the central limit theorem we expect the beam to spread out into a gaussian profile as a result of multiple scattering,

$$P(\theta) \sim e^{-\theta^2/2\langle\theta^2\rangle} \qquad (10)$$

Because of all the approximations we have made, this probability function under-estimates the number of particles scattered through large angles, greater than $3\langle\theta^2\rangle^{\frac{1}{2}}$. Thus the formula works well for about 99% of the incident beam. All the approximations seem very rough, but they affect only what we put in the logarithm; the final answer does not change much if we make rough approximations here.

Ionization

Multiple scattering does not create a beam profile signal; indeed its main effect is to blow up the profile. However, we may modify our Coulomb scattering cross section to estimate the number of target atoms ionized by the passage of the beam. We will apply our Coulomb scattering cross section (eq. 1) to a glancing collision of the incident beam with a bound electron. If the incident beam momentum is p and it is scattered through an angle θ, the electron must have been given momentum $p\theta$. We will limit our consideration to collisions in which the electron remains nonrelativistic so the electron kinetic energy (hence the energy lost by the incident particle) is of the order of

$$\Delta E(\theta) = \frac{p^2\theta^2}{2m_e} \qquad (11)$$

after the collision. Here m_e is the mass of the electron. We may use the probability function for scattering through an angle θ to express the energy lost by a particle due to ionization:

$$- \frac{dE}{dx} = \int \Delta E(\theta) \; P(\theta) \; d\theta \qquad (12)$$

$$- \frac{dE}{dx} = \frac{4\pi N_e \; z^2 \; e^4}{m_e v^2} \; \ln \; (\theta_{max}/\theta_{min}) \qquad (13)$$

Where N_e is the density of electrons in the target and v is the velocity of the beam particle. Here the minimum and maximum scattering angles are determined by different physics than before; the maximum scattering angle may be set somewhat arbitrarily, roughly equal to the maximum ionization energy of interest. About 5 kilovolts is reasonable. A heuristic argument that sets the scale for a mininum scattering angle is based on the time scale of the collision as compared to the time it takes a bound electron to move around its nucleus. If the collision force acts over a time long compared to the electron's Schrodinger frequency, the electron moves towards and away from the colliding particle many times during the collision. This greatly reduces the net work done on the bound electron by the collision force, so the electron is not kicked free of the atom. The electron Schrodinger frequency v is connected to the binding energy by $E = hv$. The duration of the interaction should be of the order of the impact parameter r divided by the particle velocity v. Lorentz contraction causes the electromagnetic field lines to be compressed into a pancake around the particle, so the duration of the collision is further shortened by the contraction factor γ. We must sum this scattering cross section over all the electrons in the target, accounting for the binding energy and Schrodinger frequency of each electron. A commonly quoted result based on the Thomas-Fermi model of the atom is the Bethe-Block formula:

$$- \frac{dE}{dx} = \frac{4\pi \; N_e \; z^2 \; e^4}{Vol \; m_e \; (\beta c)^2} \; \ln \left[\frac{2 \; m_e \beta^2 \; c^2}{IZ \; (1-\beta^2)} \right] - \Sigma_i c_i - \delta \quad (14)$$

In this formula we recognize the logarithm as coming from the integral (eq. 13). The argument of the logarithm includes the factor

$$\gamma = 1/(1 - \beta^2)^{\frac{1}{2}}$$

that comes from the maximum duration of interaction that results in an ionization. It also includes I, the mean ionization energy that determines the Schrodinger frequency of the electron. Reasonable results are obtained when one puts $I = 12$ eV. The additional terms on the right side are relatively small corrections: $\Sigma_i c_i$ reduces the energy loss to account for electrons too tightly bound to be considered in this approximation, and the term σ accounts for shadowing in dense targets.

The energy loss rate is large for incident particles of low velocity, due to the $(\beta c)^{-2}$ factor in the formula. At high energies v approaches the velocity of light and the $1/\gamma^2$ term causes a logarithmic increase in the energy loss rate. The unlimited rise of the energy deposition indicated in the Bethe-Block formula is not realistic. It appears in this formula because the rate of energy loss by very energetic particles undergoing very small deflections has been overestimated; the formula neglects dielectric shielding by neighboring atoms. The competition between the $(\beta c)^{-2}$ term and the logarithm is responsible for the notion of a "minimum ionizing" regime, the broad minimum in the energy deposition versus particle energy. For singly charged particles, the minimum loss rate is 1-1.5 MeV gram^{-1} cm^2.

The kinetic energy of the ejected electron must also be considered; sufficiently energetic electrons go on to ionize other atoms in the target. These electrons are termed δ rays. We may compute the energy loss by starting from the expression for the scattering probability (eq. 2) and substituting for the kinetic energy T of the knock-on electron:

$$T = \frac{p^2 \theta^2}{2 m_e} \tag{15}$$

We find

$$p(E) dE = 2\pi N_e z^2 e^4 \frac{dE}{E^2} \tag{16}$$

Bremsstrahlung is a particularly important energy loss mechanism for electrons. It is the result of abrupt acceleration of the incident electron by the Coulomb potential of the target nuclei. The energy loss of an electron in matter is expressed as

$$dE/dx = E/L_r \tag{17}$$

where L_r, the radiation length, is a function of the charge density of the target material.

Since multiple scattering, pair production and bremsstrahlung all begin with scattering of the incident particle by the Coulomb force of the target nuclei, these processes are all expressible in terms of the radiation length.

$$\text{Pair Production} \equiv \sigma_{pair} \simeq \frac{7}{9} \ (L_r N / Vol)^{-1} \tag{18}$$

Detection Hardware: Faraday Cup

So far we have discussed the processes by which the beam produces or liberates charged particles in a target medium. If the target medium is thick enough to completely stop the beam and all the charges it has created or liberated in the target, all these mechanisms can be ignored; the beam's kinetic energy goes into heating the target, and the net charge on the target increases by the charge of the stopped beam. We can measure the deposited charge with an electrometer or charge-sensitive amplifier, with no worries about nonlinearity. Devices of this sort are called Faraday cups. Signal charges of the order of picocolulombs or picoamperes can be measured practically. However, the requirement that the beam and all its byproducts be stopped in the detector means that targets with transverse dimensions of the order of a few radiation lengths (centimeters) must be used. This is a serious limit on resolution for many applications. Otis, et al.[6] describe an array of Faraday cups with resolution of approximately 3 mm, used on a 50 MeV proton beam. A mask is used in front of the array of Faraday cups to improve spatial resolution. This mask is a variation on the "pepper pot" described in Sander's lecture.

Faraday cups collect the entire charge in the beam, but kinetic energy of the beam goes into heating the Faraday cup instead of into the observed signal. We really only need a fraction of the beam's energy to produce a useful signal. Many detectors have been designed to measure the secondary emission current produced by the beam. In contrast to the Faraday cup we need some of the beam's scattered byproducts to leave the target. Since secondary emission electrons typically have energies up to a few kilovolts, they are unlikely to escape from inside a dense medium. Therefore the target should either be a thin wire or a gas.

K. Budal's study of charge emission from a solid target[7] describes the secondary emission efficiency of various metal plates and wires, and some essential characteristics of the signal. Budal distinguishes between δ rays (~ 1 KeV electrons kicked out by the particle beam) and secondary emission electrons, themselves produced by scattering of δ rays. The distinction is evident in the dependence of signal current on bias voltage between the target and its surroundings. If the target is negatively biased, the emission current increases significantly with the magnitude of the voltage. The negative bias increases the number of very low energy secondary emission electrons that escape from the target; without the bias voltage they would scatter inside the target and be recaptured by the target atoms. A positive bias has the opposite effect on the signal but to a lesser degree. The positive bias dramatically reduces the contribution of secondary emission electrons to the signal. Since a large fraction of the δ rays produced by the primary beam have energies in the kilovolt range, positive bias voltages of a few hundred volts do not reduce the signal current very much.

For fixed target geometry, Budhal found the signal current to be a linear function of beam current to about 2-3% precision. However, the signal has a highly nonlinear dependence on target thickness. The probability that knock-on electrons liberated inside a thick target scatter and recombine depends strongly on the target geometry and thickness. Budal measured the signal produced by passing a 10-20 GeV/c proton beam through rods and plates of copper, aluminum and berylium. He obtained 1-10 signal electrons per proton, depending on target geometry. The signal increased about 30% when the beam energy was doubled, for a given plate material and geometry.

One might expect that the target must be in vacuum to work properly; Budal found that a target coated with an insulator such as aluminum oxide will give a good signal even if it is in air.

A wire target, Figures 2a and 2b, can be made thin enough to provide a nondestructive profile measurement of a beam in a synchrotron. Carbon or tungsten filaments about 30 microns thick have been scanned through the beam at CERN.[8] Wire diameters as small as 4 microns have been used in SLC.[9] At CERN, position transducers are used to give the wire position to 5 micron accuracy. Because the

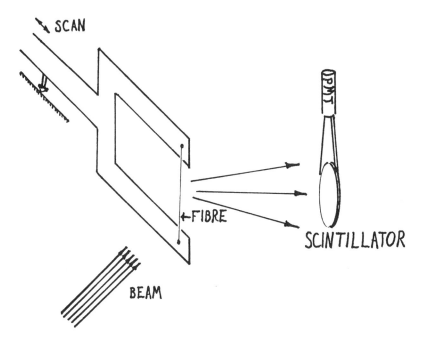

Figure 2a. Scanning wire fibers are typically 5-25 microns. The scan resolution is comparable to the fiber.

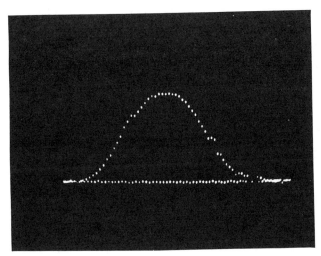

Figure 2b.

SINGLE WIRE SCAN. POSITION HEBT   6
2.6 CM FWHH (0.5 MM/STEP)

target is so thin, the secondary emission products are detected by intercepting them in a scintillator connected to a photomultiplier. The signal particles have very high energy and are detected in scintillators located well downstream of the wire and outside the beampipe. Two sets of detectors make possible the measurement of counter-rotating beams. The SPS profile monitor has 4 micron precision in its position readout. The wire moves only 7 microns per revolution of the SPS beam, yielding high-precision measurements of beams with 120 micron sigmas. It is routinely used with stored beams in the intensity range 10^9 - 6 x 10^{11} particles. The relatively slow scan speed implies rather high energy deposition in the wire; one might expect it to be destroyed by heating. Bosser, et al. calculated a 5000° C temperature rise for the wire monitor in the SPS. Fortunately most of the deposited energy is transported away from the wire in the form of kinetic energy of the secondary emission electrons, so the wire is not destroyed. Gannon, et al.[10] describe a similar system in the Tevatron. In this application, the photomultiplier signals are gated so that profiles of individual bunches can be measured.

To observe a beam in a linac, the entire profile should be acquired from a single linac pulse. One way to achieve this is to place an array of wires (Figures 3a and 3b) in the path of the beam, and measure the secondary emission current from each wire. Such a detector is described in a paper by R. Chehab, et al.[11] and is in use on the 1 GeV LAL-Orsay linac. This detector consists of an array of 0.1 mm tungsten wires separated by 0.2 mm, each connected to an operational integrator with 1 picoamp typical leakage current. The wires are placed near to a positively biased (< 100V) ground plane made of 0.1 mm aluminum foil. The signal charge was found to be about 3% of the incident beam charge. The integrator outputs are multiplexed into an amplifier so the sequence of readings can be displayed as an oscilloscope trace. This device was used on beams from 30-1000 MeV, for profiles down to 0.3 mm.

Collection of Ions

One need not place a wire in the path of the beam to produce a detectable signal. There is always some residual gas in the path of the beam, and the ions produced by the passage of the beam can be collected (Fig. 4) and detected. W.H. DeLuca[12] published measured ion currents for a beam passing through 10^{-8} Torr residual gas. Only 10^{-8} ion pairs per proton per centimeter of path in the gas are

Figure 3a. Secondary Emission Monitor grid - AGS linac.

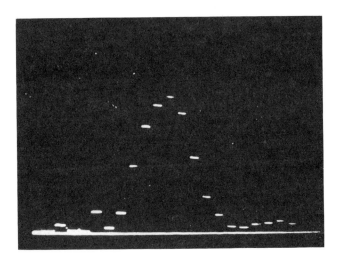

Figure 3b.

MULTI WIRE MONITOR POSITION HEBT   6
2.6 CM FWHH (5 MM/CHANNEL)

produced. He describes several approaches for detecting and displaying the ionization current. One example, the Argonne ZGS beam viewer, accelerated the ionization electrons into a phosor coated screen. This device could produce usable images from beams of 10^8 protons. Another ZGS profile monitor measured the current produced by the ions as they were collected on an array of wires, each with its own charge-coupled amplifier. These currents were only 10 nanoamperes per centimeter of pathlength in the gas for a 300mA proton beam. High accelerating voltage and crossed magnetic fields can be used to prevent the ions from drifting transversely and producing a blurred image. The positive ions produced in gas scattering have rather low kinetic energy, typically 0.014 eV. This is roughly equal to their thermal energy at room temperature. Thus the positive ions are kept aligned more effectively by the collection voltage, and blurring can be kept around ± 0.3 mm. Ionization and secondary electrons are produced with kinetic energies 10 eV and up, so they can drift transversely before collection.

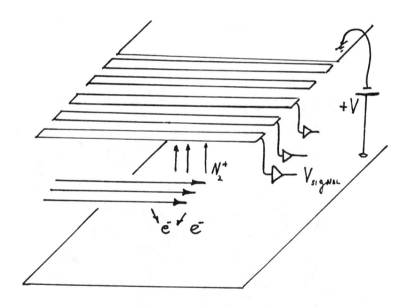

Figure 4. Residual Gas Ionization Profile Monitor.

Channelplate arrays have a current gain of about 10^6. John Krider describes their use to multiply the ion current signal in a gas ionization drift chamber.[13] In this way high resolution images have been obtained for a 1 microamp beam in 10^{-7} Torr vacuum. The channelplates are degraded by high current; Krider found their gain is reduced 16% after collecting 0.74 Coulombs/cm^2.

Fluorescent Flags

Observation of beam profiles on fluorescent screens is perhaps the oldest and simplest accelerator diagnostic technique. Used with a modern TV camera and straightforward computer image processing, fluorescent flags offer high resolution and reliability. Resolution is limited by grain size of the fluorescent material, about 30 microns.

At SLC, a chromium-doped alumina (Al_2O_3:Cr^{3+}) phosphor is used to measure the beam profile. The phosphor can be made by anodizing aluminum in an electrolyte of potassium dichromate and sodium tetraborate, in a process described by R.W. Allison, et al.[14] He describes an entertaining acoustic "scintillation" or popping phenomenon that occurs during the process, and is associated with the accretion of the chromium ions. The light output is red, 693 and 694 nanometer wavelengths. Allision, et al. measured no degradation in light output after exposure to 8×10^{15} 5 GeV protons or 5×10^{15} electrons at 3.4 MeV. Yencho and Walz[15] report that the image produced by 10^8 electrons can be seen on a vidicon TV camera, and the linear range extends beyond 10^{12} electrons. They quote a resolution of 50-100 microns.

M.C. Ross, et al.[16,17] describe similar performance for a phosphor commonly used in oscilloscopes, Gd_2O_2S:Tb. It is a powder and can be bonded to an aluminum substrate with barium silicate. Radiation damage of such a screen is evident for exposures of about 5×10^{15} electrons in a 0.5 mm spot. For exposures 2/3 of this amount, no damage was observed. The phosphor responds linearly up to 10^{10} electrons in a 100 micron spot.

Zinc sulfide phosphor is still used for imaging beams. Neet[18] reports its sensitivity to be 10^{10} protons per cm^2; the phosphor is rendered inactive after exposure to 10^{17} protons per cm². Variants of ZnS are the oscilloscope phosphors P-3 and P-31. They may be used to image an X-ray beam as well as a particle beam. Neet lists the scintillation properties of other materials, such as sapphire, quartz and lithium glass.

Cerenkov Light

Typically the light output from a fluorescent screen will decay in milliseconds. Information on the bunch structure of the particle beam cannot be extracted from the light output. Cerenkov radiation, however, is prompt and its intensity is proportional to the instantaneous beam current. This made it possible to measure the profile of any one of a train of very closely spaced electron bunches in the Los Alamos free electron laser experiment.[19]

The velocity of light in a medium with index of refraction n is c/n. Cerenkov light is emitted by a charged particle moving faster than light can propagate in the medium. The Cerenkov light, Fig. 5, propagates away from the particle beam making an angle θ.

$$\cos(\theta) = c/(vn) \qquad (19)$$

where c is the velocity of light in a vacuum and v is the velocity of the emitting particle. This velocity dependence of the emission angle has been used to discriminate the energies of low velocity particles. The number of photons of frequency ν emitted per centimeter of path in the refractive medium is given by

$$\frac{d^2N}{dxd\nu} = \frac{2\pi z^2 \sin^2(\theta)}{137\ c} \qquad (20)$$

The detailed chemistry of the medium does not affect the light output. Sheffield, et al. describe the use of quartz screens to image single beam pulse from an electron linac. The light from a single bunch was selected using a gated image intensifier. Quartz was chosen because it is very resistant to radiation damage. For very high current beams, it is possible to use a flowing liquid medium as the source of Cerkenkov light.[21]

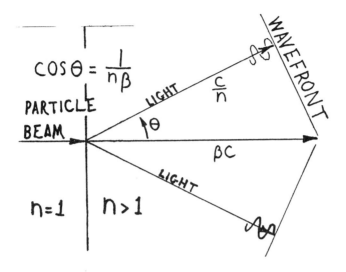

Figure 5. Cerenkov Light Index of Refraction n > 1

$$\beta = (v/c) > (c/n)$$

Transition Radiation

Transition radiation is emitted by a particle when crossing a boundary between media of different dielectric constant. This type of radiation offers an interesting way to measure the energy of the particle; the angle of emission of the light is a strong function of the energy of the particle, in contrast with Cerenkov radiation. It has been used for simultaneous measurement of profile and divergence of electron beams at Los Alamos.[2 2]

Consider a particle passing through a substance with a complex refractive index ϵ with $|\epsilon| > 1$. The intensity of the transition radiation with angular frequency ω emitted by the particle as it exits the medium is a function of the particle's energy and the direction of its velocity.

$$\left[\frac{d^2 W}{d\omega \, d\Omega} \right]_{Fwd} = \frac{e^2}{\eta^2 c} \frac{\alpha^2}{(\alpha^2 + \gamma^{-2})^2} ; \quad |\epsilon| > 1 \qquad (21)$$

The medium boundary can be tilted at any angle without affecting the emission angle. Radiation emitted in the backward direction has the same angle dependence as forward radiation; however, the direction for which $\alpha = 0$ is the direction that light colinear with the particle beam would be reflected by the medium boundary. The intensity of the backward radiation is modified by a Fresnel term characterizing the reflectivity of the medium boundary.

$$\left[\frac{d^2W}{d\omega\, d\Omega} \right]_{Back} = F(\Psi, \alpha, \omega) \left[\frac{d^2W}{d\omega\, d\Omega} \right]_{Fwd} \tag{22}$$

Wartski, et al. (Fig. 6) made use of the forward and backward radiation from two thin films of aluminum to produce an interferometer, for which the angle dependence of the intensity is an explicit function of the film L separation and the energy $\hbar\omega$ of the observed light:

$$\tag{23}$$

$$\frac{d^2W}{d\omega\, d\Omega} = 4F(\Psi, \alpha, \omega) \frac{e^2}{\eta^2 c (\alpha^2 + \gamma^{-2})^2} \sin^2 \left(\frac{\pi L}{2\lambda} (\gamma^{-2} + \alpha^2) \right)$$

Fiorito, et al[22] made use of the polarization of the transition radiation from a Wartski interferometer to deduce the divergence of the electron beam in the Los Alamos free electron laser linac. The beam energy could be determined with about 1% precision from the intensity of the transition radiation.

Synchrotron Radiation

Synchrotron radiation[23,24] has provided useful diagnostic signals for electron and proton beams. At the energies of SSC and LHC, synchrotron radiation techniques will be readily applicable.

Electromagnetic radiation, from radio waves to X-rays, is produced by accelerating or decelerating charges. The power spectrum of the radiation is proportional to the power spectrum of the acceleration of the charges. In a synchrotron, the acceleration comes from steering the particles along curved

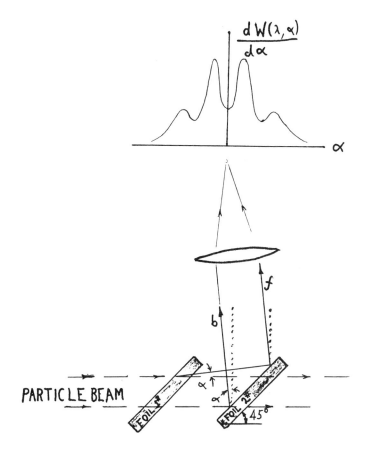

Figure 6. Transition Radiation from a Wartski Interferometer. The "forward" transition radiation (f) produced at foil #1 is specularly reflected from foil #2. It interferes with "backward" transition radiation (b) produced on foil #2. The top of the figure shows a typical interference pattern described by Eq. 23.

trajectories in the bending magnets. The energy lost by an electron following a circular path of radius r is

$$E(KeV) = \frac{88.5 \ E^4 \ (GeV)}{r(meters)} \tag{24}$$

The radiation has a broad spectrum in wavelength; the power midpoint of the spectrum is called the critical wavelength:

$$\lambda_c (Angstroms) = \frac{18.64}{B(Tesla) \ E^2 (GeV)} \tag{25}$$

For a horizontal band, the rms vertical opening angle of the radiation is $1/\gamma$, where γ is the ratio of total energy to rest mass. The opening angle is a function of wavelength:

$$\Psi = (\lambda/\lambda_c)^{1/3} \ 1/\gamma \tag{26}$$

One can make an image of the electron beam using the synchrotron radiation. An image of reasonable quality can be made using synchrotron radiation at visible wavelengths. The limiting resolution of such an image is set by diffraction, for a given observed wavelength.

In the NSLS VUV ring,[25] light with wavelength 400 nanometers is focussed by a simple telescope onto a one-dimensional photodiode array with 0.025 mm wide elements. Light from a bending magnet (bending radius r = 1.91 meters) at the U5 port, Fig. 7a and 7b, was used; the vertical beam size is close to its maximum here. Since the synchrotron radiation emerges in a horizontal fan, the horizontal angular acceptance of the telescope is limited to its optimum value.

$$\theta = 2 \ (\lambda/r)^{1/3} \tag{27}$$

Use of a smaller angular acceptance limits resolution due to diffraction. A larger acceptance causes light from an extended segment of the electron orbit to enter the telescope. The horizontal image thus produced would be the image of this extended source rather than a good replica of the beam cross section. The smallest beam size measured at U5 is σ_y = 60 microns. The opening angle of 400 nm light from the VUV ring is

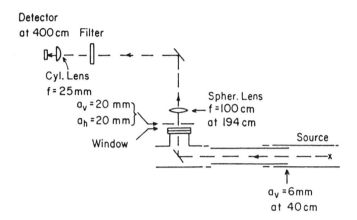

Figure 7a. Layout of optics at NSLS VUV ring beam port U5.

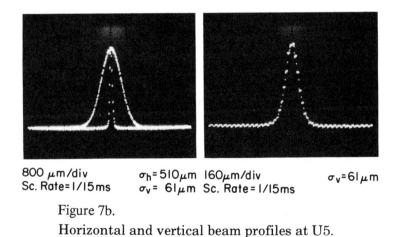

800 μm/div σ_h=510μm 160μm/div σ_v=61μm
Sc. Rate=1/15ms σ_v= 61μm Sc. Rate=1/15ms

Figure 7b.
Horizontal and vertical beam profiles at U5.

$$\Psi = \frac{0.511 \text{ MeV}}{750 \text{ MeV}} \left[\frac{400 \text{nm}}{2.5 \text{nm}} \right] = 4 \text{ milliradians}$$

We would expect a diffraction-limited resolution of

$$2.35 \ \sigma_y = 1.22 \ \lambda/\Psi$$

The right side of this expression is the familiar one-slit diffraction formula. The factor 1.22 is characteristic of diffraction by a round aperture. This formula gives the distance to the first null of the diffraction pattern, and is a pessimistic estimate of the resolution. The factor 2.35 on the left gives a more realistic estimate of σ for the image of a point source. Using a bandpass filter to avoid chromatic aberration, Nawrocky and Yu[25] tested the U5 monitor and found it produced an image with σ = 20 microns as the source size went to zero. Thermal distortions of any mirrors exposed to the full sychrotron radiation spectrum can cause errors in the measured beam position and even the beam profile. Sabersky[26] proposed an invar mirror to minimize such errors.

It is obvious that best resolution would be attained using radiation of the shortest possible wavelength. Lenses and mirrors for focussing 0.1 - 3 nanometer radiation can be used, but it is more straightforward to look at the beam through a pinhole camera. In this case the resolution limit comes from optimising two competing effects. The first effect is diffraction. The second effect is just blurring due to the fact that a pinhole really does no focussing. Neglecting diffraction, the image of a pinhole with diameter d illuminated by a point source will be

$$\Delta x = d(L_2/L_1)$$

where L_1 and L_2 are respectively the source-to-pinhole distance and the pinhole-to-obsverer distance. For a magnification of 1, $L_1 = L_2$. It turns out that the optimum pinhole size is approximately the geometric mean of the wavelength λ and the distance L from the pinhole to the source.

Hoffman[27] points out that possibly the opening angle of the radiation is such that the pinhole is not illuminated by all parts of the beam. In this case the beam size will be underestimated. To avoid such problems he suggests choosing to observe the electron beam at a point where it is converging.

Some care must be given to making the edges of the pinhole truly opaque to hard X-rays. Typically the pinhole is made from two sets of tantalum slits oriented at right angles to make a square hole. Such a pinhole is being installed at the NSLS X4 beamline. Slits 40 microns in width were electrical-discharge machined into plates of thickness 2 mm. The expected resolution is about 50 microns.

Detecting synchrotron radiation is in most cases not a small-signal problem. Usually the more important problem is protection of the detector from destruction by tens or thousands of watts of incident power. Since visible and X-ray photons can initiate ionizations and photoemissions, it is possible to detect the photon beam using all the techniques and devices for charged particle detection described earlier: scanning wires and grids or wires, gas ion chambers, and scintillating screens observed using cameras or photodiode arrays. Commercial TV cameras either viewing the synchrotron radiation directly or viewing a scintillator screen will give useful images for beam currents of the order of microamperes.

Acknowledgement

The author is grateful to V. Castillo for collaborating on the preparation of the manuscript and to Mrs. Joan Depken for typing the final document.

References

1.   David M. Ritson, Techniques of High Energy Physics, New York, Interscience Publishers, Inc. pp. 1-53 (1961).

2.   John D. Jackson, Classical Electrodynamics, second edition, New York, John Wiley and Sons, pp. 618-683 (1975).

3.   Robert B. Leighton, Principles of Modern Physics, McGraw-Hill Book Company, New York, pp. 362-365 (1959).

4.   Walter E. Meyerhof, Elements of Nuclear physics, McGraw-Hill Book Company, New York , p. 5 (1967).

5.   M. Roos et al., ed. Review of Particle Properties, Lawrence-Berkeley Laboratory LBL-100, pp. 34-42, Revised April 1982.

6.   A. Otis, R. Larson, R. Lockey, B. DeVito, A. Van Steenbergen, "AGS Injector Beam Monitoring System," IEEE Trans. Nuc. Sc. NS-14, pp. 1116-1126 (1967).

7.   K. Budal, "Measurement of Charge Emission from Targets as a Means of Burst Intensity and Beam Intensity Monitoring," ibid., pp. 1132-1137.

8.   J. Bosser, J. Camas, L. Evans, G. Ferioli, J. Mann, O. Olsen, R. Schmidt, "The Micron Wire Scanner at the SPS," Proceedings of the 1987 IEEE Particle Accelerator Conference, E.R. Lindstrom and Louise S. Taylor, eds. IEEE Service Center, Piscataway, NJ, pp. 783-785 (1987).

9.   G. Bowden, D. Burke, C. Field, W. Koska, "Retractable Carbon Fibre Targets for Measuring Beam Profiles at the SLC Collision Point," Nuc. Inst. Meth. A-278 (3). pp. 664-669 (1989).

10.   J. Gannon, C. Crawford, D. Finley, R. Flora, T. Groves, M. MacPherson, "Flying Wires at Fermilab," proceedings of the 1989 IEEE Particle Accelerator Conference, Floyd Bennet and Joyce Kopta, eds., IEEE Service Center, Piscataway, NJ, pp. 68-70 (1989).

11.   R. Chehab, J. Bonnard, G. Humbert, B. LeBlond, J. L. Saury, "A Multiwire Secondary Emission Profile Monitor for Small Emittance Beams," IEEE Trans. Nuc. Sci. NS-32, pp. 1953-1955 (1985).

12.   William H. DeLuca, "Beam Detection Using Residual Gas Ionization," IEEE Trans. Nuc. Sci. NS-16, pp. 813-822 (1969).

13.   John Krider, "Residual Gas Beam Profile Monitor," Nuc. Inst. Meth. A278(3), pp. 660-663 (1989).

14. R.W. Allison, Jr., R.W. Brokloff, R.L. McLaughlin, R.M. Richter, M. Tekawa, J.R. Woodyard, "A Radiation-Resistant Chromium-Activated Aluminum Oxide Scintillator," Lawrence Radiation Laboratory UCRL-19270 (1969).

15. S.Y. Yencho, D.R. Walz, "A High Resolution Phosphor Screen Beam Profile Monitor," IEEE Trans. Nuc. Sci. NS-32, pp. 2009-2011 (1985).

16. M.C. Ross, J.T. Seeman, R.K. Jobe, J.C. Sheppard, R.F. Steining, "High Resolution Beam Profile Monitors in the SLC," ibid pp. 2003-2005.

17. M.C. Ross, N. Phinney, G. Quickfall, H. Shoaee, J.C. Sheppard, "Automated Emittance Measurements in the SLC," Proceedings of the 1987 IEEE Particle Accelerator Conference, IEEE Service Center, Piscataway, NJ, pp. 725-728 (1987).

18. D.A. G. Neet, "Beam Profile Montiros for Fast and Slow Extracted Proton Beams," IEEE Trans. Nuc. Sci. NS-16, pp. 914-918 (1969).

19. R.L. Sheffield, W.E. Stein, R.W. Warren, J.S. Fraser, A. Lumpkin, "Electron Beam Diagnostics and Results for the Los Alamos Free Electron Laser," IEEE Journal of Quantum Electronics QE-21 (7) 895-903 (1985).

20. David M. Ritson, Techniques of High Energy Physics, Interscience Publishers Inc., New York, pp. 318-320 (1961).

21. M. Buttram, R. Hamil, "Cherenkov Light as a Current Density Diagnostic for Large Area, Relatively Pulsed Electron Beams," IEEE Trans. Nuc. Sci. NS-30, pp. 2216-2218 (1983).

22. R.B. Fiorito, D.W. Rule, A.H. Lumpkin, R.B. Feldman, D.W. Feldman, B.E. Carlsten, "Optical Transition Radiation Measurements in the Los Alamos Free Electron Laser Driver," Proceedings of the 1989 IEEE Particle Accelerator Conference, Floyd Bennet and Joyce Kopta, eds.., IEEE Service Center, Piscataway, NJ, pp. 65-67 (1989).

23.   K. Hubner, "Synchrotron Radiation," in Proceedings of the CERN Accelerator School-Synchrotron Radiation and Free Electron Lasers, S. Turner, ed. CERN 90-03, pp. 24-52 (April 3, 1990).

24.   A. Hofmann, "Electron and Proton Diagnostics with Synchrotron Radiation," IEEE Trans. Nuc. Sci. NS-28, pp. 2132-2136 (1981).

25.   R.J. Nawrocky, J. Galayda, L.H. Yu, D.M. Shu, "A Beam Profile Monitor for the NSLS VUV Ring Employing Linear Photodiode Arrays," IEEE Trans. Nuc. Sci. NS-32, pp. 1893-1895 (1985).

26.   A.P. Sabersky, "Monitoring the beams in SPEAR with Synchrotron Light," IEEE Trans. Nucl Sci. NS-20, 638-641 (1973).

27.   A. Hofmann, K.W. Robinson, "Measurement of Cross Section of a High-Energy Electron Beam by Means of the X-Ray Portion of the Synchrotron Radiation," IEEE Trans. Nuc. Sci NS-18.

LONGITUDINAL EMITTANCE

AN INTRODUCTION TO THE CONCEPT AND SURVEY OF MEASUREMENT TECHNIQUES INCLUDING DESIGN OF A WALL CURRENT MONITOR

Robert C. Webber
Fermi National Accelerator, Batavia, Il. 60510

ABSTRACT

The properties of charged particle beams associated with the distribution of the particles in energy and in time can be grouped together under the category of longitudinal emittance. This article is intended to provide an intuitive introduction to the concepts of longitudinal emittance; to provide an incomplete survey of methods used to measure this emittance and the related properties of bunch length and momentum spread; and to describe the detailed design of a 6 Ghz bandwidth resistive wall current monitor useful for measuring bunch shapes of moderate to high intensity beams. Overall, the article is intended to be broad in scope, in most cases deferring details to cited original papers.

INTRODUCTION TO LONGITUDINAL EMITTANCE

The focus of the present discussion is to be longitudinal properties of beams of accelerated particles. These properties include the energy, energy distribution, and time distribution of particles comprising a beam. Longitudinal emittance will be defined as a concept to quantify the distribution, in energy and time, of particles in a beam. Since accelerating fields are generally periodic, a phase variable often replaces the time variable.

Beams from even dc types of accelerators, such as ion sources, electron guns, Van de Graaffs, Cockcroft-Waltons, etc., usually display a time structure due to pulsed operation, random fluctuations in the beam current, or both. While these time structures are of interest to the designers and users of such beams, the dimension of time is not fundamental to the acceleration process. This clearly is not the case with accelerators utilizing periodically pulsed or rf fields to accelerate particles. There are only certain phases and amplitudes of the fields appropriate for stable particle acceleration. Time or phase becomes a fundamental parameter.

Consider Figure 1A. The positions of many particles, say from an ion source or electron gun, are plotted on a scale of energy and time. E_s has arbitrarily been defined to be the average energy of the distribution. Particles of the beam have some range of energies, but are distributed more or less uniformly in time. The energy spread implies a spread in velocities. The particles, therefore, move relative to each other. Arrows in the figure indicate this relative motion.

Now accelerate this beam with a sinusoidally varying field as in Figure 1B. Figure 1C results. The peak to peak energy spread has increased, but the time distribution is still uniform. Observed some time later, without additional acceleration, higher energy particles are seen to arrive early in time relative to lower energy companions as shown in Figure 1D. The beam has become bunched in time.

The motion of several particles can be followed through this process. Note that the phase coordinate has been chosen to move with the unaccelerated particle S, which therefore

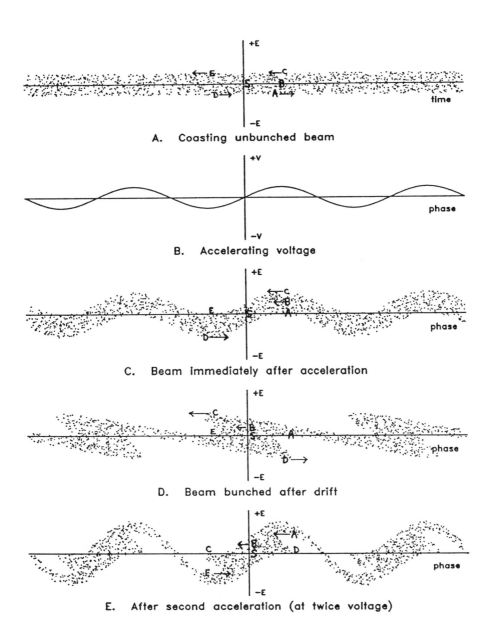

A. Coasting unbunched beam

B. Accelerating voltage

C. Beam immediately after acceleration

D. Beam bunched after drift

E. After second acceleration (at twice voltage)

FIGURE 1
BEAM DISTRIBUTIONS IN PHASE SPACE

remains fixed in both energy and phase. Particles **A**, **B**, and **C** are accelerated equal amounts. Particle **A** is accelerated to the same energy as particle S and then remains fixed in position. It undergoes motion on the energy coordinate, but not on the phase coordinate. **B** is accelerated to a higher energy than S and proceeds to travel earlier in time. **C** changes in energy the same amount as **A** and **B**, but, being the most energetic to begin with, moves most in phase relative to S. **E** and **D** are decelerated the same amount as **A**, **B**, and **C** are accelerated. **E** moves to an energy equal to that of S, mimicking **A**. **D** ends up as low in energy, relative to S, as **C** does high; and thus, proceeds to lag behind as much as **C** leads.

If this beam is left to drift indefinitely, it can be seen that **B**, **C**, and **D** will move arbitrarily far from S. The original bunch will disperse and new ones will periodically appear and fade away.

As drawn, S is special in that the phase coordinate is chosen to move with it and in that it remains un-accelerated. In the initial distribution, however, S was not special at all, except that it (and many other particles) happened to be at the energy chosen as the reference energy. It remained un-accelerated by our arbitrary choice of rf phase. We are free to choose any particle to be our reference particle. S became special by choice. We may even choose a particle which, unlike S up to now, is accelerated. The coordinate system must then simply be redefined so that the energy coordinate, like the time or phase coordinate, tracks with the energy of the chosen reference. The coordinate system is usually defined so as to move with the ideal particle of the beam. That particle, called the synchronous particle, always has the design energy and has a phase relative to the accelerating field so as to maintain that design energy. The energy and time or phase coordinate system, moving with the synchronous particle is called phase space. The coordinates are then E-E_s, energy relative to the synchronous particle, and ϕ-ϕ_s, phase relative to the synchronous particle. Particle motion in phase space is then motion relative to the ideal particle. The synchronous particle remains fixed in phase space, changing in energy and advancing in time exactly as desired, simply because that is how it has been defined. The area occupied by the beam particles on the phase space plane is the longitudinal emittance. Usually we are just concerned with the area occupied by one bunch, since we deal with a periodic system. Typical units for longitudinal emittance, which has dimensions of energy times time, are eV-seconds. In some instances it is useful to transform the vertical axis to momentum, instead of using energy.

As the beam bunch was left in Figure 1D, any particle not on the horizontal axis is destined to move arbitrarily far from S, the reference. This presents no problem if the beam is intentionally left to coast unbunched or if it is on its way to an experiment concerned only with the energy, not the time structure, of the beam. But our intent is to accelerate the beam.

Assume for the moment that S is accelerating. In order that the whole beam accelerate along with it, all particles must "stay close" to S in our phase plane coordinate system. If allowed to move far from S, they are being allowed to become much different from S in time and/or energy; exactly what we do not desire. Nevertheless, with a non-zero energy spread, we cannot completely freeze the motion of all particles relative to S. The beam has a finite emittance, and phase and energy are coupled. The success of our efforts, the stability of the beam, then relies on developing a system to constrain motion to a bounded region of phase space near S.

A hint of the possibility of a stable solution to this problem can be found by imagining the bunched beam in Figure 1D to again be exposed to an accelerating field. If that field is still phased so S sees zero field with a positive slope, it is seen that the higher energy particle C is decelerated, slowing or even reversing its motion away from S. The lower energy particle D is accelerated, bringing its energy closer to that of S and slowing or reversing its motion away from S. Continuing this process, it will be observed that it is possible for particles to travel in "orbits" around S as the drifting and acceleration continues. This action whereby the energy of all particles is focused toward that of the synchronous particle is known as longitudinal phase focusing. The frequency with which a particle "orbits" around the synchronous particle is the synchrotron oscillation frequency. In a circular machine, the number of oscillations per turn (usually much less than one) is the synchrotron tune.

The strength of this focusing action determines the amplitude and frequency of the oscillatory motion relative to the synchronous particle. The amplitude is a measure of how far in energy or time a stable particle is allowed to travel. The focusing strength is determined by the slope of the accelerating field (the difference in acceleration seen by the synchronous particle relative to another particle at a different phase) and by the rate of time or phase slip between the synchronous particle and another particle with a different energy.

The situation described so far, requires particles with energies greater than that of the synchronous particle to "arrive early", and particles with lesser energies to "arrive late". This is indeed the case for linear accelerators and straight beamlines, since the path traveled by particles is essentially energy independent. In circular machines, the situation is complicated by the fact that, in any given bending field, higher energy particles are bent less than lower energy particles. The higher energy particles, though traveling faster than lower energy particles, follow an orbit with a larger average radius and thus travel a longer path. Which arrives first? At ultra high energies, velocity is essentially the speed of light for all particles independent of small energy differences. In this case, the higher energy particles, having farther to go, actually arrive late. Just the opposite of the situation we started with. At much lower energies the path length and velocity factors compete; and, for each machine, there is some energy for which the effects exactly cancel, the time of arrival is independent of energy. This energy is called the transition energy. Below this energy, higher energy particles do arrive early with respect to lower energy particles.

Remembering that the longitudinal focusing action relied on decelerating higher energy particles and accelerating lower energy particles relative to the synchronous particle, it is noted that the sign of the slope of the accelerating field must be reversed above transition when higher energy particles begin arriving late. This in fact is necessary in machines which accelerate beam through the transition energy. It is accomplished by shifting the phase of the rf by $180° - 2\phi_s$, where ϕ_s is the phase of the synchronous particle. The transition energy of a given machine for a given type of particle, E_t, is determined by the circumference and the quadrupole strengths. This energy is usually referred to in terms of the relativistic parameter γ, where $\gamma_t = E_t/m_o$. γ_t is approximately equal to ν_h, the horizontal betatron tune. For example, in the Fermilab Main Ring, γ_t is about 18.7.

The requirement to change the slope of the accelerating field at transition means that the longitudinal focusing strength is changing sign there. For the sign to change (since the focusing strength changes in a continuous manner), the magnitude must pass through zero. The sign change is effectively compensated by shifting rf phase; but the magnitude of the phase focusing strength cannot be prevented from dropping to zero at transition. Since this focusing effect is what bounds motion in the phase space, its disappearance opens the possibility for unbounded motion. Particles are free to move arbitrarily far in time or energy relative to the synchronous particle. Fortunately, any motion in this phase space has a finite velocity, so particles remain near S for some time. In any case, if transition cannot be avoided, loitering in its vicinity should be.

Having constructed the concepts of longitudinal phase space and longitudinal emittance as potentially useful ways of looking at particle beams, how do we relate them to measurable and controllable quantities? There are numerous treatments of longitudinal phase space and motion with any depth of mathematical rigor desired.[1,2,3,4] This paper will stick close to the final results of those efforts without worrying about the derivations beyond the simple intuitive description.

The focusing strength, dependent on the slope of the accelerating field, is a function of the magnitude of a sinusoidally varying rf field of a chosen frequency. It also depends on the difference in time of arrival for particles of different energies, i.e. the drift length (or equivalent) between accelerating cells or the radius of a circular machine. These are real quantities with physical limits for real machines. Finite longitudinal focusing strength implies a finite area of our phase space in which bounded motion relative to the synchronous particle is possible. The obvious limiting case, already discussed, is that in which the beam simply drifts, the rf voltage is zero. The area of bounded motion in phase space, called the bucket area, is then identically zero.

The bucket area also depends on the rate of acceleration. Everything else being equal, the area of a "moving bucket", one in which the synchronous particle accelerates, is less than the area of a "stationary bucket" in which the synchronous particle maintains constant energy. Intuitively, a given rf field has only a limited capacity for affecting a particle's energy. If some of this capacity is used to cause an average energy change, the capacity for longitudinal focusing is reduced.

For a circular machine, the bucket area is given by

$$S = \frac{8\beta}{2\pi f_{rf}} \sqrt{\frac{2E_s V}{\pi h |\eta|}} \, \alpha(\Gamma) \quad \text{eV-sec} \tag{1}$$

where β is the relativistic velocity factor; f_{rf}, the rf frequency; E_s, the energy of the synchronous particle in eV; V, the peak rf amplitude in volts; h, the harmonic number (ratio of rf frequency to revolution frequency); and $\eta = (1/\gamma_t^2 - 1/\gamma^2)$, the transition effect factor. γ is simply the particle's total energy normalized to its rest energy. At transition γ equals γ_t and η is identically zero; the bucket area becomes infinite. $\alpha(\Gamma)$ is the moving bucket factor which modifies the bucket area as the synchronous phase is changed. For a stationary bucket, $\phi_s = 0°$, $\alpha(\Gamma) = 1.0$, for $\phi_s = 20°$ $\alpha(\Gamma) \simeq 0.5$, and for $\phi_s = 90°$ $\alpha(\Gamma) = 0.0$. The bucket area depends on the accelerating field amplitude, $S \propto V^{1/2}$, and on the difference in arrival times of particles with slightly different energies, $S \propto (E_s/|\eta|)^{1/2}$.

The maximum momentum spread between the synchronous particle and a particle at the edge of the bucket is

$$\frac{\Delta p}{p} = \pm \left[\frac{1}{cp} \right] \sqrt{\frac{2E_s V}{\pi h |\eta|}} \, \beta(\Gamma) \tag{2}$$

where cp is the particle momentum in eV units, and $\beta(\Gamma)$ is another "moving bucket" factor depending on the synchronous phase. $\beta(\Gamma) = 1.0$ for $\phi_s = 0.0$, $\simeq 0.5$ for $\phi_s = 35°$, and 0.0 for $\phi_s = 90°$. References 4 and 5 contain graphs and tabulations of $\alpha(\Gamma)$ and $\beta(\Gamma)$.

Particles need not, and frequently do not, occupy the entire bucket area. The bunch area, i.e. the longitudinal emittance, for stable particles is less than or equal to the bucket area. The actual portion of the bucket occupied depends on the energy and time distributions of the particles in the beam. It is these quantities which we must attempt to measure in order to quantify the longitudinal emittance of a real beam. In terms of the time (phase) spread, which can often be measured, the longitudinal emittance is found to be approximately[4]

$$\epsilon_L = \frac{\pi}{16} SQ^2 (1 - \frac{5Q^2}{96}) \sqrt{\cos(\phi_s)} \quad eV-sec \qquad (3)$$

where S is bucket area in eV-sec and Q is the peak phase excursion of any particle relative to ϕ_s. Q is 1/2 the total phase spread, in radians, (peak bunch length) of the beam. This expression is accurate to about 6% for beam filling a stationary bucket. The approximation deteriorates for large synchronous phases and for large amplitude motions, i.e. bunch areas comparable to the bucket area.

The rate at which particles "orbit" around the synchronous particle in phase space, the synchrotron frequency, is given by

$$f_s = \frac{f_{rf}}{\beta} \sqrt{\frac{|\eta| V}{2\pi h E_s} \cos(\phi_s)} \left[1 - \frac{Q^2}{16} \right] \quad hertz \qquad (4)$$

where all quantities have been previously defined and E_s has units of eV. The last term accounts for the fact that the frequency is somewhat amplitude dependent; that term is, in fact, only an approximation, correct for small amplitudes. Notice that the η term is contained in the numerator, forcing the oscillation frequency to zero at transition. Note also that bucket area, emittance, and synchrotron frequency all scale as the square root of the rf voltage.

It should be realized that no real particle need be identical to the ideal synchronous particle. Particle distribution in phase space is not necessarily uniform nor symmetric about the synchronous particle. Figure 2 shows some possible distributions in phase space and the projection of those distributions on the time axis. The synchronous particle is at the origin of each plot and the buckets are assumed stationary. Evolution of each distribution is followed for one half synchrotron period. The first distribution, uniform and symmetric around the synchronous particle, appears stationary though individual particles circulate around within. The second distribution represents a bunch whose centroid begins with a phase or timing error. The time projection maintains a uniform shape, but oscillates back and forth relative to the synchronous phase. The last distribution has an asymmetric aspect ratio such that at one phase of the motion the particles are grouped tightly in phase with a large

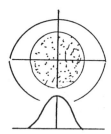 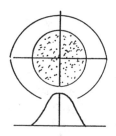 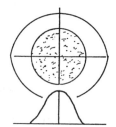

A.   Symmetric and centered

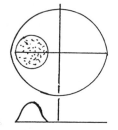 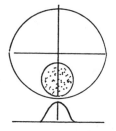 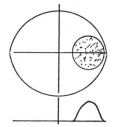

B.   Symmetric and off-center

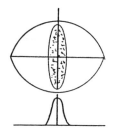 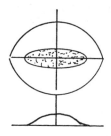 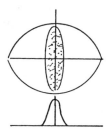

C.   Asymmetric and centered

FIGURE 2
POSSIBLE PARTICLE DISTRIBUTIONS
AND THEIR EVOLUTION IN A BUCKET

energy spread, and 90° later the energy distribution is reduced at the expense of increased time spread. Any combination of these and higher order mode distributions are possible.[6]

Bucket area must be sufficiently large to contain all particles of interest at any phase of their motion in order to maintain stability. The emittance of the beam must be considered as the entire phase space area covered and/or enclosed by the particles during one complete oscillation, not just the area they happen to occupy at any instant of time. The last two distributions in Figure 2 are examples of beam mismatched to the bucket. Such mismatches are generally undesirable because the effective momentum spread of the beam is increased and loss of particles may result. Mismatches result in an inefficient use of bucket area since more rf voltage, and therefore more power and money, is required to contain the beam than would be necessary in a matched condition.

The time projections of the phase space beam distributions begin to take on real significance to the engineer when it is realized that what is being considered is a time distribution of charged particles moving through space. This is simply an electric current which can be measured to yield the time or phase distribution of the particles in the beam.

MEASUREMENT METHODS

Techniques of measuring particle energies and time distribution rely on only a few fundamental physical principles, including the radiation of and interaction with electric and magnetic fields by charged particles, and the relationship between velocity and energy. The ways in which these principles have been exploited are innumerable. Methods used depend on many variables -- What is the availability of the beam? Single pass or circulating? May the measurement be destructive of the beam? What is the energy range of the particles? What time resolution is required? What spatial resolution is required? What type of particle is to be measured? What is the end use of the information? And so on... A few examples will be presented.

ENERGY MEASUREMENT

THE SPECTROMETER

The fundamental spectrometer technique relies on the ability to subject the beam to energy dependent forces and monitor subsequent motion. A classical spectrometer is sketched in Figure 3. An example exists at the end of the Fermilab 200 MeV Linac. All or a portion of the beam is directed through a magnet, which introduces a bend angle[5]

$$\theta = \frac{L}{\rho} \tag{5}$$

where L is the magnet length, and ρ is the radius of curvature of the particle path in the field B.
Since

$$B\rho = \frac{p}{q} \tag{6}$$

find

$$\theta = \frac{BLq}{p} \tag{7}$$

where p is the particle momentum and q its charge.

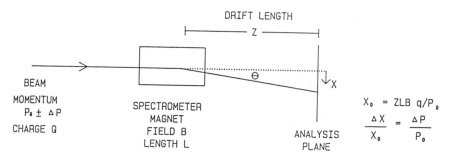

FIGURE 3
TYPICAL SPECTROMETER SETUP

Some distance z downstream, a particle of momentum p will be displaced by

$$x = \frac{zBLq}{p} \tag{8}$$

The spread in energy of a beam with a central momentum p_o is found from

$$\Delta x = \frac{-zBLq}{p_o} \frac{\Delta p}{p_o} = -x_o \frac{\Delta p}{p_o} \tag{9}$$

$$-\frac{\Delta x}{x_o} = \frac{\Delta p}{p_o} \tag{10}$$

The fractional displacement spread is equal to the fractional momentum spread.

Note that this treatment assumes that all particles enter the spectrometer magnet with the same angle and at the same position, that is, it assumes zero transverse emittance. A real beam has finite transverse emittance. The contribution of this emittance to the position spread at the observation point must be accounted for to obtain a correct momentum spread measurement. Either, the momentum sensitivity of the position is made so large as to overwhelm position spread due to transverse beam size; the measurement is made at one location for different known transverse focusing conditions; or, the the beam is observed at more than one spot in a beamline with known optics so that the transverse effect can be calculated out.

At Fermilab the beam is observed at one location using a single wire which may be scanned in position across the beam. At any wire location, the charge deposited on the wire by the beam by secondary emission effects is electronically integrated to provide a signal proportional to the amount of beam intercepted. A histogram of beam density vs. position thus generated can be interpreted as a momentum distribution plot.

Circular accelerators or storage rings are effectively continuous spectrometers. Any ring has a dispersion function[2], determined by the transverse focusing optics, of some value at any location. The dispersion function, D, relates the radius of the particle orbit to its momentum, $\Delta R = D(\Delta p/p)$. This effect has been used in the Fermilab

Tevatron to measure momentum spread and calculate longitudinal emittance. The horizontal beam profile is measured at two locations where the dispersion is different and the transverse lattice functions are known. These profiles contain the information necessary to de-convolve transverse and longitudinal emittance contributions. Beam profiles are measured with "flying wire" monitors[6]. Another aspect of the "continuous spectrometer" appears in frequency domain measurements when the effect of momentum on revolution time is considered. This aspect will be discussed later.

A spectrometer scheme is also used to measure the energy spread of the e^+ and e^- bunches in the SLAC Linear Collider.[7,8,9,10,11,12] The magnet which splits the electrons and positrons into their respective transport lines to the collision point is used as the spectrometer magnet. The energy spread of the beam and the dispersion introduced by this horizontal bend combine to produce a spread in the horizontal beam size at some distance downstream where the beam passes through the vertical wiggler magnet to induce synchrotron radiation. The horizontal width of the radiation source is the same as the beam size and therefore directly related by the dispersion to the momentum spread. The synchrotron radiation shines on a phosphor screen viewed by a TV camera. The TV image is digitized and processed to produce an energy histogram. Separate systems measure each of the two beams, positrons and electrons. The absolute energy of the beams is measured by sensing their centroid positions in the high dispersion regions using stripline beam position monitors. The results of this measurement provide important real-time feedback to the linac to stabilize the energy.

TIME OF FLIGHT MEASUREMENTS

The energy of any particle may be determined from its velocity if its rest mass is known. The velocity can be calculated if the time required to travel between two points along a known path is measured. Energy determinations can be made quite sensitively using this method for particles with velocities much less than $\beta = 1$. As particles become relativistic the slope of the velocity vs. energy curve approaches zero, as does the sensitivity of this energy measurement technique. Two examples of this technique are presented.

The Fermilab Linac is in the midst of an upgrade project to replace the four high energy tanks with new side-coupled accelerator structures to increase the Linac output energy from 200 MeV to 400 MeV. To insure optimal performance, the energy gain per cell and the phasing between cells must be correctly adjusted. A method of measuring the energy gain as a function of rf amplitude and phase, based on schemes from Los Alamos[13,14], has been implemented on several existing linac tanks in preparation for this effort. The beam comes in a 30μsec pulse bunched by the 200 Mhz rf. Beam signals are available from stripline pick-ups located at each end of each tank. Signals from the upstream and downstream end of the tank of concern are compared by an electronic phase detector circuit. Phase comparison of signals of a known frequency is equivalent to a time measurement. As the rf amplitude and phase is adjusted, the phase between the upstream and downstream beam signals can be monitored. Interpreted as a change in time of flight between the two striplines, the energy gain of the beam within the tank can be computed. This method has proven successful and has already demonstrated the existence of matching errors between tanks in the existing linac.

H$^-$ beams from the ATS (Accelerator Test Stand) radio frequency quadrupole (RFQ) accelerator at Los Alamos are analyzed in energy using time of flight techniques.[15,16] Laser

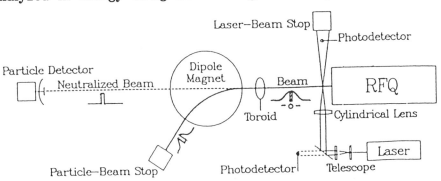

FIGURE 4
LOS ALAMOS ACCELERATOR TEST STAND
LASER-INDUCED NEUTRALIZATION DIAGNOSTIC APPARATUS
(figure taken from ref. 15)

induced neutralization is used as a means of sampling a short time segment of an H⁻ bunch from the RFQ. See Figure 4. A 32 picosecond laser pulse is focused to a 30 micron diameter and shined on the beam, perpendicular to the direction of beam travel. The laser energy strips one electron from the H⁻ ions, creating a neutral beam from a 200 micron slice of the bunch's longitudinal dimension. A dipole magnet sweeps away remaining charged particles while the neutral segment travels straight ahead to a fast secondary emission monitor (SEM). Detection of the laser light by a fast photodetector initiates the timing interval which terminates when beam arrives at the SEM. For the known drift length and measured time interval, the energy of the particles can be calculated. The timing of the laser pulse is measured relative to the phase of the RFQ rf. This information and the ability to sample only a small fraction of the bunch length allows measurement of particle energy as a function of rf phase. Integration of the signal from the neutral particle detector provides the information necessary to determine particle density vs. phase. It is thus possible to measure the complete longitudinal phase space parameters of the beam.

BUNCH SHAPE AND TIME DISTRIBUTION MEASUREMENTS

The Los Alamos laser induced neutralization scheme demonstrates that particle time distribution can be determined by arranging to tag a small time slice of the bunch and sample only the tagged particles. Few situations allow the slick technique of altering the charge state of a slice of beam, but other implementations of sampling techniques are possible. These methods usually rely on the transformation of temporal distribution to spatial distribution.

Two such examples involve processing the beam of secondary electrons generated from a wire or foil target intercepting the beam to be measured. A method developed at Brookhaven[17], Figure 5, modulates an acceleration voltage applied to the secondary electron beam in synchronism with the rf of the primary beam. The energy modulated electron beam travels through a region of uniform magnetic field. The momentum spread of the electron beam is transformed to a spatial spread on a plane where each electron has

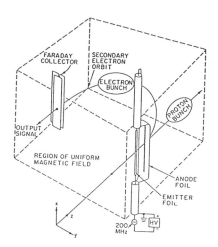

FIGURE 5

BROOKHAVEN BUNCH LENGTH MONITOR

(figure taken from ref. 17)

completed one half of a circular path in the magnetic field. A Faraday cup charge collector with a narrow entrance slit placed on this plane provides a signal from only those secondary electrons with the "right" energy, that is, those due to particles in the primary beam at a time when the electron accelerating voltage is at a particular value. By varying the phase of the electron accelerating voltage relative to the bunch, the "right" accelerating voltage is moved in time relative to the bunch. Thus, the charge measured at the Faraday cup is a representation of a particular time sample of the original beam.

A modification of this scheme, developed at INR in the USSR, Figure 6, accelerates all secondary electrons to the same energy.[18] This electron beam is then passed between a pair of deflecting plates driven with an rf voltage synchronous with the beam rf. The position of these particles some distance downstream becomes a function of the time at which they were deflected. Since this beam has the same temporal distribution as the primary beam, that distribution is effectively transformed to a spatial distribution, the position of which may be translated by adjusting the phase of the deflecting voltage. Again, a Faraday charge collector with a narrow entrance slit samples a small part of the spatial distribution to provide an electrical signal. Time resolution the order of 1° at 600Mhz is reported. Bunch shape

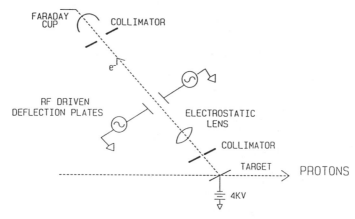

FIGURE 6
USSR INR BUNCH LENGTH
MONITOR SCHEME

measurements by this device under different conditions of the primary beam have allowed calculation of longitudinal emittance values for the beam.

Bunch shape measurements have been made at the Daresbury Synchrotron Radiation Source using synchrotron light focused onto an image dissector tube.[19]   The front end of the image dissector tube consists of a photocathode. Photoelectrons produced there have essentially the same time distribution as both the synchrotron light and the original electron beam which is desired to be measured.   In the image dissector tube, the photoelectrons are accelerated and operated on by a transverse deflection scheme similar to that just described.   The Faraday cup is replaced by an electron multiplier dynode structure with a slit entrance aperture. Transit time spread of electrons through the multiplier limits the time resolution of this device.

At SLAC, Cherenkov light from a radiator positioned to intercept the electron beam is focused onto a streak camera.[20] The streak camera is another device utilizing photoemission and fast electron deflection to transform a high speed time image to a spatial image which can be electronically processed.   Streak cameras are also used at Orsay to measure time duration of picosecond electron bunches.[21]

Where time resolutions of picoseconds are not required,

various types of current transformers are generally used for bunch shape measurements.[22,23,24,25]   One such type of transformer is the "wall current monitor".[26,27,28]   This type of device can offer bandwidths of several gigahertz.   A wall current monitor is described in detail later in this paper.

Applied to beams in beamlines or linacs, current monitor signals are generally processed by oscilloscopes (which still rely on the ability to transform time into position by the deflection of electrons).   These may be conventional analog scopes, sampling scopes, or one of the growing number of high speed digitizing sampling scopes.   Where the sampling rates are too slow in real time, interlaced sampling using external triggering synchronized to the accelerator rf allows a bunch profile signal to be built up over many pulses.   Thus the full risetime bandwidth of the sampling head may be realized, subject to any jitter in the triggers.   These techniques are, of course, applicable also to signals from current monitors in circular machines.[29]   All result in direct time domain representations of bunch shapes.

An ever increasing number of video and image processing techniques and technologies are available for capturing, archiving, and analyzing bunch shape information that can be displayed on oscilloscope screens.   These include video frame grabbers, video image tape recording, etc.   For instance, beam bunch signals, as viewed on a scope screen by a video camera, are routinely recorded by a standard VCR during antiproton acceleration cycles and transfers when Fermilab operates in the collider mode.   This provides a record in the event it becomes necessary to autopsy a doomed antiproton injection or acceleration cycle.

Beam bunch signals from circular accelerators are frequently presented in the form of mountain range displays, e.g. Figure 7.   Such displays are generated by adding the beam current signal to a slowly increasing voltage ramp or staircase.   The resulting signal is displayed on an oscilloscope triggered synchronously with the accelerator rf.   Multiple triggers during the slow voltage ramp, result in a family of traces offset in vertical position on the scope screen.   Thus, the time evolution of a fast bunch signal can be monitored and displayed in a single picture.   The delay between each trace of the family can be made arbitrarily long by adjusting the repetition rate of the rf synchronous triggers and the slope of the slow ramp.   This is simplified with a two channel scope that allows addition of the signals from the

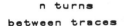

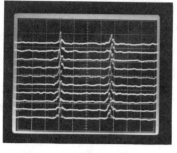

time →

FIGURE 7
TYPICAL MOUNTAIN RANGE DISPLAY

two channels.   This type of display facilitates the observation of bunch phase oscillations (synchrotron oscillations of the centroid of the bunch), bunch shape oscillations, and special bunch manipulations such as bunch coalescing used in the Fermilab collider.   Even intrabunch momentum oscillations can be observed if the difference signal from a position monitor in a high dispersion region is viewed on this type of display.[30]

If the signal from a fast beam current monitor is used to trigger a sample of the machine rf waveform, bunch phase oscillations can be measured without the need for a mountain range display or wideband scope.   This circuit is essentially a type of phase detector.   Variations of the sampled output indicate beam motion in time relative to the rf.

Bunch shape oscillations of the type shown in the third beam distribution of Figure 2 are easily observed by peak detection (e.g. by a diode) of a fast beam current signal. The resultant signal may be observed on low bandwidth equipment.   Signals processed in this manner have been used to provide stabilizing feedback to beams in synchrotrons.

FREQUENCY DOMAIN INFORMATION

In recent years much attention has been given to frequency domain interpretation of signals from beam monitors in circular machines.[30,31,32]   Several treatments with varying degrees of mathematical rigor are available.   A few results and applications will be mentioned here.

An unbunched beam circulating in a synchrotron or storage ring will produce signals in a current monitor at all multiples of the revolution frequency.   This is a consequence of the quantum nature of the particles and their random distribution around the machine.   In this sense, the signal is like the shot noise associated with, for instance, currents in an electron tube.   These random signals are called Schottky signals and are proportional to the square root of the number of particles.   Since particles of different momenta in the unbunched beam are free to circulate at correspondingly different revolution frequencies, each makes a signal contribution at its own frequency.   Consequently, the spectrum of the Schottky signal has a characteristic width at each revolution harmonic directly proportional to the momentum spread of the beam and the harmonic number. This effect allows rather easy determination of the energy spread of a circulating beam.   The signal from a current monitor is simply displayed on a spectrum analyzer. Scaling is

$$\frac{\Delta f}{f} = -\eta \frac{\Delta p}{p} \tag{11}$$

Schottky signal currents are quite weak since they scale with the square root of the number of particles, rather than linearly.   For this reason, beam pickups to sense the Schottky signals are generally designed to be relatively narrowband tuned pickups, as opposed to wideband devices like the wall current monitor.   Wideband devices typically present only a few ohms impedance to the beam, whereas tuned devices are easily constructed to present the beam with kilo-ohms.   Since signal power is proportional to this impedance, the advantage is obvious.   The momentum spread of beams in antiproton accumulator rings is usually measured by this technique just described.[33]   Longitudinal stochastic cooling relies on Schottky signals.[34]

Bunched beams, on the other hand, by definition consist of particles which on the average have exactly the same revolution frequency as all other particles in the beam.   This average frequency is controlled by the rf system which maintains the bunch by its phase focusing action.   A particle, which may at one instant of time have a higher frequency than the synchronous particle, must at some later time, have a correspondingly lower frequency in order to remain part of the bunch. The result is that the bunch now creates coherent signals with a predictable time structure, i.e. the pulse in time we recognize as the bunch shape. If the bunch shape is known, the spectrum of the periodic bunch signal can be calculated as a Fourier series without resorting to statistical methods as is necessary for the Schottky signal.

The Schottky signals do not, of course, cease to exist. The beam still consists of quantized charges with some randomness of distribution within the bunch. The longitudinal Schottky signals appear as fm sidebands around each harmonic of the revolution frequency. The revolution harmonic is arbitrarily narrow for fixed rf frequency, while each fm sideband signal has a finite width related to the momentum spread. The appearance of fm type sidebands due to the variation of individual particle revolution frequencies around the synchronous frequency can be appreciated if compared to an rf communication signal in which the instantaneous signal frequency is varied around a central frequency. The fm sideband components are displaced from the corresponding revolution harmonic signal by multiples of the synchrotron oscillation frequency. Since the synchrotron frequency is generally small compared to the revolution frequency, the fractional frequency separation between the strong coherent revolution harmonic signal and the weak incoherent sideband is very small. Recovery of longitudinal information contained in an fm sideband requires a specially designed narrowband receiver. There are usually easier ways to acquire desired information from bunched beams.

One such method to derive bunch length information by processing coherent frequency domain signals from a beam current monitor or appropriate stripline monitor is used at Fermilab.[35] The technique begins with the assumption of an appropriate bunch shape model, in this case gaussian. The Fourier signal components required to represent such a bunch, periodic at the revolution frequency, are then computed as a function of the rms bunch length. The sensitivity of different spectral components to changes in bunch length is determined and two components with adequate sensitivity are selected. In the Fermilab application, the selected frequencies are the first and third harmonics of the rf frequency. Frequency selective receivers detect the amplitude of these components and an analog circuit computes the bunch length. The bunch length is proportional to the square root of the natural logarithm of the ratio of the two spectral amplitudes. A real time bunch length signal, with 10 Khz bandwidth, is thus generated for use as a beam feedback signal or simply as a diagnostic tool.

This survey represents only a few of the methods that have been used to get a handle on the quantitative aspects of particle beams related to their longitudinal emittance -- energy, energy spread, and bunch shape. In addition to the many references cited here, the Particle Accelerator School publications offer introductory descriptions to many topics and the Particle Accelerator Conference Proceedings provide a wealth of technical detail.

A WALL CURRENT MONITOR

Knowledge of the time distribution of particles in a beam is important for assessing the performance of an accelerator and for monitoring its operation. To this end, a device for measuring beam current with flat amplitude response and the best possible time resolution is essential. The words "broadband current transformer" could generically describe a device with the desired characteristics. A simple current transformer is depicted below in Figure 8.

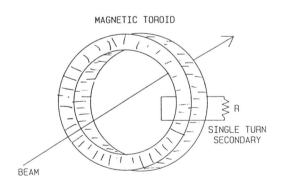

FIGURE 8
SIMPLE CURRENT TRANSFORMER

The response of this device is easily calculated observing that the magnetic flux in the core is equal to the sum of that resulting from the beam current and that resulting from the current in the secondary winding. Assuming a single turn secondary, we have

$$\Phi_T = \Phi_b + \Phi_s = Li_b + Li_s \tag{12}$$

where L is the inductance of any single turn winding on the magnetic toroidal core.

$$L = \frac{\mu\mu h}{2\pi} \ln \frac{b}{a} \tag{13}$$

where h is the core thickness and a and b are respectively the inner and outer diameters of the core.

Faraday's law of induction tells us that the induced current i_s will be of a polarity so as to oppose the flux due to i_b. Noting this and that the induced voltage in the secondary is equal to the rate of change of total flux we write

$$\frac{d\Phi_T}{dt} = v_s = L \frac{di_b}{dt} - L \frac{di_s}{dt} \tag{14}$$

Substituting $v_s = i_s R$ and switching to frequency domain notation we find

$$V_s(\omega) = j\omega L I_b(\omega) - j\omega L \frac{V_s(\omega)}{R} \tag{15}$$

Solving for V_s

$$V_s(\omega) = j\omega L \frac{I_b(\omega)}{1+j\omega L/R} \tag{16}$$

This is the response of a high pass filter with a 3db point at $\omega_c = R/L$ and a passband transfer function $v_s = i_b R$.

The transformer effectively forces a current equal to the beam current through the load resistor. The current is measured by measuring the voltage across the load. This is good for frequencies above $f_c = \omega_c/2\pi = R/2\pi L$. Notice that there has been no requirement for any "wall" currents in any beam tubes and the low frequency corner can be made arbitrarily small.

Now, change the geometry of the original picture without changing the effective circuit. Rotate the secondary winding so the resistor appears inside the core, then revolve the whole secondary about the longitudinal axis of symmetry. The core is now surrounded by a metal housing except for a uniform resistive gap in the inner diameter. A cross section is shown in Figure 9.

A beam tube may be added to each end of the transformer without changing the circuit response, which is determined solely by the gap resistance and the inductance presented by the core. The beam tube can even close upon itself and have numerous grounds without affecting the signal because these paths are now simply in parallel with the arbitrarily low resistance of the housing. See Figure 10. The housing also shields the resistive gap from external signal sources. At dc, the attenuation of external signals is simply the ratio of the housing resistance to the gap resistance. For ac signals, the attenuation is further enhanced, as the inductance presented by the core appears in series with the gap resistance. Finally, at rf frequencies, the skin effect prevents any external signal from reaching the gap.

One further mechanical modification

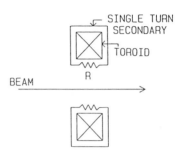

FIGURE 9
SIMPLE TRANSFORMER
NEW GEOMETRY

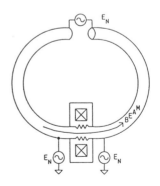

FIGURE 10
TRANSFORMER IN
ACCELERATOR

and the addition of a gap voltage monitor tap completes the initial conceptual design of the monitor. The mechanical modification is the addition of a ceramic vacuum seal between the beam and the gap resistance, allowing the resistance and the magnetic material to reside outside the vacuum. The ceramic is not necessary if the vacuum properties of the material are adequate for the application. However, the trend toward ultrahigh vacuums, especially in storage rings, makes the use of the ceramic seal generally advisable. Half the cross section now appears as shown in Figure 11.

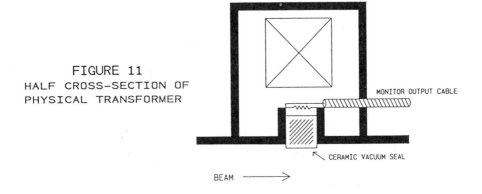

FIGURE 11
HALF CROSS-SECTION OF
PHYSICAL TRANSFORMER

A beam current transformer has thus been developed, but at least two questions remain open -- What happened to the wall current monitor concept, and what limits the high frequency response?

The electric and magnetic fields due to the beam are properly thought of as electromagnetic waves propagating out from the beam. At low frequencies, the magnetic component of the wave penetrates the metallic walls of the beam tube and the transformer. This is how the flux can appear in the magnetic core of our transformer. In fact, the dc component of a beam current totally enclosed within a conducting nonmagnetic pipe can be measured external to the pipe without the need for any type of insulating break in the pipe. The dc magnetic field exists external to the pipe and can be measured, for example, with a second harmonic type dc transformer.

At higher frequencies, the wave is attenuated as it propagates through the conducting beam tube -- the skin effect. At high enough frequencies, where the skin depth is a fraction of the wall thickness, the attenuation is effectively complete. No high frequency field appears external to the beam tube. The only way for the fields due to the beam currents to not appear in some region of space is for them to be exactly canceled by equal and opposite fields due to some other current distribution. Since the beam fields appear within the beam tube, but not outside of it, the cancelling fields must be generated by currents in the wall of the tube. At last, the long lost wall currents! So the idea of a wall current monitor is seen to be correct at frequencies where the skin depth in the wall is small. The idea of wall currents can also be introduced intuitively by considering that a bunch of beam electrically induces an opposite charge in the beam tube. As the beam bunch moves, so too must the induced wall charge; hence, wall currents.

To see these currents, it must be arranged for the resistor to appear in series with the wall. What foresight, that's just where it was last left in Figure 11. The effect of wall currents equal and opposite to the beam flowing through the resistor is identical to the effect of the low frequency transformer forcing a current equal and opposite to the beam through the resistor. There is no "cross-over" problem, only a change in point of view.

The idea of wall currents allows consideration of bandwidth limitations of the monitor in terms of circuit models. The impedance through which this wall current must flow as it passes the monitor is the parallel combination of the impedances presented by the physical gap, the intended resistance with its strays, and the surrounding volume loaded with magnetic and possibly other material. The voltage to be

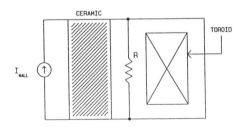

FIGURE 12
WALL CURRENT MONITOR
CIRCUIT MODEL

measured and related to the beam current is that due to the portion of the wall current which flows in the intended resistance. Therefore, it must be insured that all or at least a constant fraction of the wall current follows that path rather than a parallel path. The circuit can be modeled as in Figure 12. It looks amazingly like the mechanical sketch of Figure 11 rotated 90°. Keeping in mind the goal of insuring that a large constant fraction of the current flow through the resistor, the requirements of each component can be studied.

Begin with the cylindrical outer volume, partially filled with magnetic material. It forms a cylindrical cavity, presenting some shunt impedance across the resistor. This impedance ought to be made large relative to the resistance at all frequencies of interest; if not ignorably large, at least relatively constant. At low frequencies, the impedance is simply the single turn winding inductance, chosen to set the low frequency L/R time constant. At higher frequencies, determined by the magnetic material, the cavity exhibits appreciable loss. The shunt inductance begins to look more like a resistor; still OK if the value is large compared to the gap resistance. The shunt impedance of the cavity at moderate frequencies increases linearly with the number of cores. This number may be increased as space permits. At some high frequency, electromagnetic waves impinging on the surface of the material will not penetrate; they will be reflected with little attenuation. At this point the material does little good except to define the boundary of a now smaller cavity which may resonate and present a low impedance across the resistance at some frequencies. Different additional material may be added to handle frequencies in this range.

This graded approach was chosen for our design, filling the volume with several different materials. Farthest from the gap are Ceramic Magnetics, Inc., MN60, manganese-zinc ferrite cores with $\mu \simeq 6000$. Four of these cores with dimensions 4" i.d., 6" o.d., and 1" high provide a low frequency inductance of about $40\mu H$ and a broadband resistance of $\simeq 20\Omega$ in the ten to hundred Mhz range. Closer to the gap, two Ceramic Magnetics C-2025 ferrite cores, $\mu \simeq 175$, provide loss to higher frequency signals, attenuating them before reaching the MN60 which becomes less useful at these frequencies. Closer yet to the gap are two epoxy cores loaded with microwave absorbing material. The remaining volume near the gap is filled with a foam type microwave absorbing material. The inner wall of the cavity is coated with a microwave absorbing paint to further damp any

surface currents induced by waves managing to find their way
to the walls.   Indeed, the object is to make this volume
appear as a black box into which energy enters, as into a
high impedance, and is totally absorbed.   Our design results
in a broadband impedance of between 10 and 20 ohms up to
multi-gigahertz frequencies.   This effective cavity shunt
impedance dictates a choice of around one ohm for the gap
resistance, in order to limit the shunting effect to about 10%.

Satisfied with
minimizing the cavity
effect, the circuit
simplifies to that shown
in Figure 13.
Pessimistically, the
ceramic section can be
modeled as a lumped
capacitance which sets the
high frequency rolloff of
the device at $\omega = 1/RC$.
Taking a more optimistic
point of view, the ceramic
can be seen to form a
radial transmission line (a
distributed rather than a
lumped element circuit)

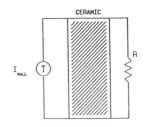

FIGURE 13
WALL CURRENT MONITOR
SIMPLIFIED CIRCUIT MODEL

through which currents from the inside of the beam tube
must flow to reach the monitoring resistor.   If the resistor
terminates this line in its characteristic impedance there is
theoretically no frequency sensitivity.

To estimate the characteristic impedance of this radial
line, look from the inside of the beam tube radially outward
and unroll the cylinder to a plane. The gap then simply
looks like a parallel plate transmission line with a plate
separation equal to the gap width and plate width equal to
2π times the radius at which the ceramic resides.   The
impedance of such a line is

$$Z = 377 \; \frac{h}{w} \; \frac{1}{\sqrt{\epsilon_r}} = 377 \; \frac{h}{2\pi R} \; \frac{1}{\sqrt{\epsilon_r}} \tag{17}$$

where ϵ_r is the dielectric constant of the ceramic filling the
volume between the plates. For our gap, $r \simeq 1.75$", $h =$
0.125", and $\epsilon_r \simeq 9.5$.   The radial line impedance is found to
be 1.4Ω.   This ceramic section was made to our specifications

to match the previously determined magnitude of the gap resistance, about one ohm. The radial extent of this line is basically the wall thickness of the ceramic, in our case, 3/16". The high dielectric constant slows the radial wave propagation, effectively making the line look longer by $\sqrt{\epsilon_r} = 3$. The circuit now is as shown in Figure 14.

FIGURE 14

TRANSMISSION LINE MODEL

Some validity to assuming the radial line point of view is lent by the data in the Figure 15. It shows measurements made on a wall current monitor of a previous design, exhibiting a ≃20db peaking in the frequency response. In an attempt to explain this undesirable response, the radial transmission line model was applied. The radial line impedance for that ceramic geometry was calculated to be approximately 3.2 ohms. The resistance of the device was 0.32 ohms. The length of the radial line corresponded to one quarter wavelength at 1.9 Ghz, the same frequency at which the peaking was observed. Response of a Spice program model of a current source driving that transmission line and termination agrees with the measured resistor voltage over the full range of measurements.

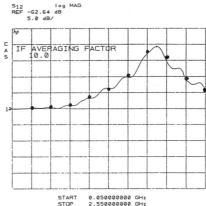

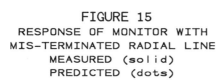

FIGURE 15

RESPONSE OF MONITOR WITH
MIS-TERMINATED RADIAL LINE
MEASURED (solid)
PREDICTED (dots)

With an understanding of the surrounding cavity shunt impedance and the radial transmission line model in hand, consideration of the resistance itself and connection of a suitable voltage monitor remain. A distributed resistance is appropriate since other parts of the monitor are treated as distributed transmission lines. We use 80 120Ω rf chip resistors evenly spaced around the circumference. These resistors are used because they exhibit very high self-resonant frequencies, i.e. they really act as constant value resistors over the frequency range of concern. The resistors are soldered to metal bands on either end of the ceramic gap and are kept close to the outer surface of the ceramic. This provides approximately the required 1.4Ω resistance with good high frequency characteristics.

Design of a voltage monitor requires an understanding of the voltage distribution around the circumference of the gap. This voltage may properly be considered as the linear superposition of two contributions: 1) a sum or average gap voltage (the zeroth order term in a harmonic expansion of the azimuthal voltage distribution), and 2) a difference voltage (all higher order terms in the harmonic expansion).

Only energy propagating outward through the radial transmission line contributes to the sum voltage. The difference voltage, on the other hand, results only from energy propagating azimuthally around the gap. It is important to realize that these two paths are different from each other, with grossly different propagation constants. The radial line, properly terminated, will transmit the sum signal with little attenuation, producing a sum signal voltage at the resistors independent of frequency (above the low frequency L/R corner). To azimuthal waves, the gap looks like a parallel plate type transmission line with a plate width corresponding to the wall thickness of the ceramic and a plate separation corresponding to the gap width. This line is filled with ceramic dielectric and heavily shunt loaded by the intended gap resistance. The result is a lossy line with frequency dependent characteristics.

For centered beam, only the sum signal (i.e. the radial mode) exists, the voltage distribution is uniform by symmetry. There is no azimuthal transfer of energy and the gap voltage has no fundamental upper frequency limit. The often cited $1/2\pi RC$ limit is simply incorrect if the radial line mode is properly terminated. For off-center beam, the voltage at any point around the gap is the superposition of the frequency independent sum component and a frequency dependent difference component. The difference component sees a driving source determined by the beam position and bunch shape, and displays a frequency dependence determined by the azimuthal line parameters.

For our gap, we find azimuthal line parameters as below:

capacitance	--	3 pf/inch
inductance	--	21 nh/inch
conductance	--	0.071 mho/inch
series resistance	--	$\simeq 0 \ \Omega$

The capacitance is simply the total gap capacitance divided by the circumference. The inductance is found by considering the line without the shunting resistors. It is a line with dielectric constant $\epsilon \simeq 9$ and known capacitance per unit length, C. The propagation velocity is then $c/\sqrt{\epsilon}$, where c is the speed of light. We have

$$v = \frac{c}{\sqrt{\epsilon}} = \frac{1}{\sqrt{LC}} \tag{18}$$

$$L = \frac{1}{C} \cdot \frac{\epsilon}{c^2} \tag{19}$$

The conductance is the total gap conductance (the inverse of the gap resistance) divided by the circumference.

The line propagation constant is defined

$$\gamma = \sqrt{(R+j\omega L)(G+j\omega C)} \Rightarrow \sqrt{j\omega L (G+j\omega C)} \tag{20}$$

For $G \gg \omega C$, (f $\ll$ 3.8Ghz), we can approximate

$$\gamma \simeq \sqrt{j\omega LG} = \sqrt{\omega LG/2} \ (1+j) \tag{21}$$

Thus, difference waves propagating around the gap are attenuated at a rate of

$$e^{\sqrt{\omega LG/2}} = 19\sqrt{f \cdot 10^{-9}} \quad \text{db per inch} \tag{22}$$

At frequencies even as low as 100 Mhz, the attenuation of difference waves is 6 db per inch. The greater the attenuation, the more independent one location is from its neighbors. Each point on the circumference "knows" only that signal induced locally by the beam, since signals from some distance away are attenuated to insignificance. Therefore, any one voltage monitor point will contain significant position information. (Note, however, that the position sensitivity will be strongly frequency dependent.) Schneider[27] correctly describes a lower cutoff frequency for use of a wall current monitor as a position detector. For our monitor this occurs at 8 Mhz.

The goal of this design, however, is to provide a position insensitive beam current signal. For frequencies high enough to be strongly attenuated by the azimuthal line, the position dependent voltage around the circumference is as described by the static charge model used by Schneider. For lower frequencies, any one point on the circumference effectively "communicates" with other points, reducing the magnitude of any beam position dependence. Both Schneider's static model and calculations done at Fermilab taking the transmission line parameters into account[36] predict that proper summation of signals from four points, at 90° intervals around the gap, will produce a signal with only a few percent position sensitivity for beams within the center half of the device.

Signal summation is performed utilizing DC coupled microstrip combiners on an RT Duroid printed circuit board. Transformer type combiners were rejected because commercial devices typically do not provide the desired few kilohertz to few gigahertz bandwidth. The shunt loaded circuit shown in Figure 16 was chosen because it performs the desired summation with no frequency or source impedance dependence, provided the four input signals are identical. This constraint may seem inappropriate, since the only reason we bother to sum the signals at all is because they are not identical. However, our investigation into the response of transmission line combiners revealed that all possible configurations we considered require some compromise. For example, the typical series loaded combiner, shown in Figure 17, can display dramatic frequency dependence when the source impedance differs from the circuit line impedance, that is, if the combiner lines are not properly back terminated. This is a significant and deciding factor since the source impedance provided by the azimuthal transmission line neither

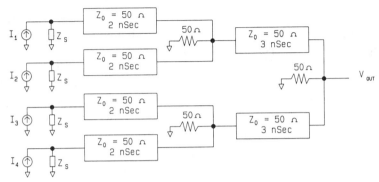

FIGURE 16
SHUNT LOADED COMBINER SCHEMATIC
as used

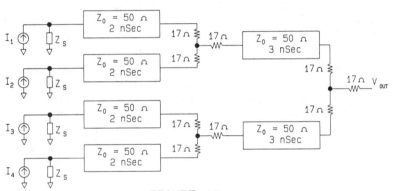

FIGURE 17
SERIES LOADED COMBINER SCHEMATIC
as rejected

matches that of the combiner lines nor remains constant with frequency. This type of combiner, in fact, produces a frequency dependence not only in the difference signal components but also in the sum signal components. Its output will be frequency dependent even when the inputs are identical, that is, when the beam is on center!

The Duroid combiner circuit board is flexible, allowing it to be physically wrapped around one edge of the monitor gap. Good connection is made with the microstrip ground plane there. The four striplines connect to the other side of the gap by small copper straps. Two 120 ohm resistors are omitted from the gap at each 50 ohm tap point to maintain approximately constant resistance per unit arc length around the circumference of the gap. The combined signal is brought out of the monitor housing on a single 0.141" semi-rigid coax from an SMA connector on the combiner board.

Figures 18 and 19 are photographs of the prototype wall current monitor.   Figure 18 shows the clamshell style outer housing, the assorted magnetic toroids, the ceramic gap spanned by many chip resistors, one of the four voltage taps, part of the combiner board, and the output connector. Figure 19 is a close-up of the gap and combiner board.

FIGURE 18
WALL CURRENT MONITOR

FIGURE 19
WALL CURRENT MONITOR
CLOSE-UP VIEW OF GAP,
RESISTORS, AND
COMBINER BOARD

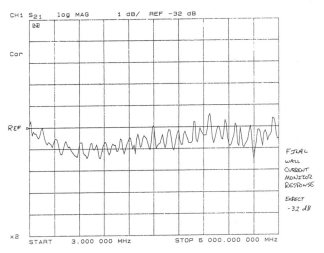

FIGURE 20
WALL CURRENT MONITOR
MEASURED FREQUENCY RESPONSE

A frequency response measurement of the final wall current monitor is shown in Figure 20. The 1db variations, apparent in the measurement at 200 Mhz intervals, are believed to be a remnant of the test fixture and not the wall current monitor itself. Frequency response, flat to within 1db over the full 6 Ghz measurement range of the network analyzer, is demonstrated. The measurement was made with a center conductor placed through the monitor to form a 50 ohm coaxial transmission line system. Tapered cones were used at the ends of the structure to match down to standard coaxial connectors and cables. An identical setup with an outer pipe minus wall current monitor was used as a reference point to which measurements were normalized. Extreme care is necessary in the mechanical design of the test fixture to assure repeatable high frequency electrical measurements each time the fixture is taken apart and reassembled.

One final point to be considered in the application of the extremely wideband wall current monitor is that, by design, it responds to currents in the vacuum chamber wall, whether or not those currents are due to the instantaneous beam current in the monitor. As short beam bunches pass steps, bellows, and other discontinuities in the vacuum chamber wall, they deposit electromagnetic energy in these volumes. The components of that energy with frequencies above the waveguide cutoff frequency of the beam tube are able to then propagate down the pipe. These waves travel at a velocity different from that of the beam and induce currents in the vacuum chamber walls as they go along. The wall current monitor, assuming it has sufficient bandwidth, will faithfully respond to these currents and produce an output signal. Such signals have been directly observed by monitors in the Fermilab Main Ring.[37] These signals, while perhaps valuable for estimating the amount of waveguide energy within the beam tube, are considered as noise when the goal is to measure beam bunch shape.

As a solution to this problem, 12 inch lengths of ceramic cylinders, loaded with broadband microwave absorbing material, are installed in the beam tube immediately upstream and downstream of the wall current monitor. These absorbers provide measured attenuation of the waveguide signals, both on the bench and with beam, of as much as 20db with no effect on the response to true beam currents. Figure 21 shows the spectral output of an earlier monitor with a single bunch circulating in the Fermilab Main Ring before and after installation of the microwave absorbers. The energy at 1.7 Ghz, above the cutoff frequency for the pipe used on that monitor, is seen to be strongly attenuated. With a single bunch, little true signal energy is present above 100 Mhz. Measurements with other beam distributions confirm that desired signals above 1 Ghz are insignificantly affected.

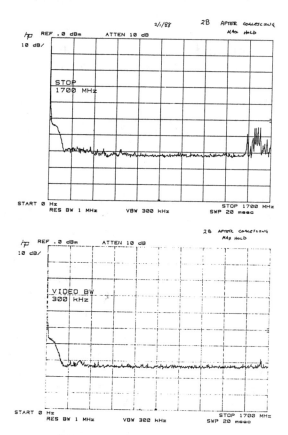

FIGURE 21
MONITOR SPECTRAL OUTPUT FOR SINGLE BEAM BUNCH
BEFORE AND AFTER INSTALLATION OF
UPSTREAM AND DOWNSTREAM MICROWAVE ABSORBERS

This concludes the description of the design parameters and response characteristics of the wall current monitor.

Bibliography

1. F.T. Cole, "Longitudinal Motion in Circular Accelerators", AIP Conf. Proc. No.153 (U.S. Particle Accelerator School), 44-82.

2. D.A. Edwards and M.J. Syphers, "An Introduction to the Physics of Particle Accelerators", AIP Conf. Proc. No.184 (1987-1988 U.S. Particle Accelerator School), 2-189.

3. W.T. Weng, "Fundamentals -- Longitudinal Motion", AIP Conf. Proc. No.184 (1987-1988 U.S. Particle Accelerator School), 243-287.

4. P.S. Martin and S. Ohnuma, "Longitudinal Phase Space in Circular Accelerators", AIP Conf. Proc. No.184 (1987-1988 U.S. Particle Accelerator School), 1941-1968.

5. C. Bovet, et.al., "A Selection of Formulae and Data Useful for the Design of A.G. Synchrotrons", CERN/MPS-SI/Int. DL/70/4, April 1970.

6. J. Gannon, et.al., "Flying Wires at Fermilab", to be published, 1989 Particle Accelerator Conference Proceedings.

7. J. Seeman and J. Sheppard, "Special SLC Developments", 1986 Linear Accelerator Conference Proceedings, September 1986, 214-219.

8. J. Seeman, et.al., "SLC Energy Spectrum Monitor Using Synchrotron Radiation", 1986 Linear Accelerator Conference Proceedings, SLAC-Report-303, September 1986, 441-443.

9. K. Bane, et.al., "Longitudinal Phase Space Measurements in the SLC Linac", to be published, 1989 Particle Accelerator Conference Proceedings.

10. E.J. Soderstrom, et.al., "Fast Energy Spectrum and Transverse Beam Profile Monitoring and Feedback Systems for the SLC Linac", to be published, 1989 Particle Accelerator Conference Proceedings.

11. C.K. Jung, "A High Resolution Synchrotron Light Detector", to be published, 1989 Particle Accelerator Conference Proceedings.

12. J. Nash, "Precision Measurements of the SLC Beam Energy", to be published, 1989 Particle Accelerator Conference Proceedings.

13. K.R. Crandall, "The Delta-T Turn-on Procedure", 1972 Linear Accelerator Conference Proceedings, Los Alamos Laboratory Report LA-5115, 122-125.

14. K.R. Crandall, "The ΔT Tuneup Procedure for the LAMPF 805 Mhz Linac", Los Alamos Laboratory Report LA-6374-MS, June 9176.

15. W.B. Cottingame, et.al., "Noninterceptive Techniques for the Measurement of Longitudinal Parameters for Intense H$^-$ Beams", IEEE Trans. Nucl. Sci., Vol. NS-32, No. 5, October 1985, 1871-1873.

16. W.B. Cottingame, et.al., "Longitudinal Emittance Measurement at the ATS", 1986 Linear Accelerator Conference Proceedings, SLAC-Report-303, September 1986, 101-103.

17. R.L. Witkover, "A Non-Destructive Bunch Length Monitor for a Proton Linear Accelerator", Nucl. Inst. and Methods, Vol.137, No.2, 1976, 203-211.

18. A. V. Feschenko and P.N. Ostroumov, "Bunch Shape Measuring Technique and Its Application for an Ion Linac Tuning", 1986 Linear Accelerator Conference Proceedings, SLAC-Report-303, September 1986, 323-327.

19. G.S. Brown, et.al., "Measurement of Bunch Length with an Image Dissector Tube", IEEE Trans. Nucl. Sci., Vol. NS-30, No. 4, August 1983, 2348-2350.

20. J.C. Sheppard, et.al., "Real time Bunch Length Measurements in the SLC Linac", IEEE Trans. Nucl. Sci., Vol. NS-32, No. 5, October 1985, 2006-2008.

21. R. Chehab, et.al., "Characterization of Low Energy Picosecond Electron Beam Pulses Using Transition Radiation", to be published, 1989 Particle Accelerator Conference Proceedings.

22. J. Borer and R. Jung, "Diagnostics", CERN Accelerator School on Antiprotons for Colliding Beam Facilities, October 1984, CERN/LEP-BI/84-14.

23. G. Lambertson, "Dynamic Devices -- Pickups and Kickers", AIP Conf. Proc. No.153 (1984-1985 U.S. Particle Accelerator School), 1413-1442.

24. T. Linnecar, "The High Frequency Longitudinal and Transverse Pick-ups Used in the SPS", CERN-SPS/ARF/78-17, August 1978.

25. T. Linnecar, "The High Frequency Longitudinal and Transverse Pick-ups in the CERN SPS Accelerator", IEEE Trans. Nucl. Sci., Vol. NS-26, No. 3, June 1979, 3409-3411.

26. R.T. Avery, et.al., "Non-Intercepting Monitor of Beam Current and Position", IEEE Trans. Nucl. Sci., Vol. NS-18, No. 3, June 1971, 920-922.

27. G.C. Schneider, "A 1.5 Ghz Wide-band Beam Position and Intensity Monitor for the Electron-Positron Accumulator (EPA)", 1987 IEEE Particle Accelerator Conference Proceedings, IEEE Catalog No. 87CH2387-9, Vol. 1, 664-666.

28. J. Sodia, et.al., "Timing Jitter Measurements at the SLC Injector", to be published, 1989 Particle Accelerator Conference Proceedings.

29. C. Moore, et.al., "Single-Bunch Intensity Monitoring System Using an Improved Wall Current Monitor", to be published, 1989 Particle Accelerator Conference Proceedings.

30. J. Gareyte, "Beam Observation and the Nature of Instabilities", AIP Conf. Proc. No.184 (1987-1988 U.S. Particle Accelerator School), 343-429.

31. R.H. Siemann, "Bunched Beam Diagnostics", AIP Conf. Proc. No.184 (1987-1988 U.S. Particle Accelerator School), 430-471.

32. R. Littauer, "Beam Instrumentation", AIP Conf. Proc. No.105 (1982 U.S. Particle Accelerator School), 869-953.

33. V.K. Bharadwaj, "Beam Transfer from the Core of the Accumulator to the Main Ring in the Fermilab Source", 1987 IEEE Particle Accelerator Conference Proceedings, IEEE Catalog No. 87CH2387-9, Vol. 2, 1022-1024.

34. J. Marriner, "Review of the Physics, Technology and Practice of Stochastic Beam Cooling", 1987 IEEE Particle Accelerator Conference Proceedings, IEEE Catalog No. 87CH2387-9, Vol. 3, 1383-1387.

35. T. Ieiri and G. Jackson, "A Main Ring Bunch Length Monitor by Detecting Two Frequency Components of the Beam", Fermilab TM-1600, June 1989.

36. J. Crisp, Fermilab, unpublished.

37. J.E. Griffin and J.A. MacLachlan, "Direct Observation of Microwaves Excited in the Fermilab Beam Pipe by Very Narrow Bunches", IEEE Trans. Nucl. Sci., Vol. NS-32, No. 5, October 1985, 2359-2361.

TRANSVERSE EMITTANCE:
ITS DEFINITION, APPLICATIONS, AND MEASUREMENT

Oscar R. Sander
Los Alamos National Laboratory
Los Alamos, New Mexico 87545

INTRODUCTION

Transverse emittance is one of the fundamental beam characteristics in any accelerating structure or beam transport line. In this paper we will first define emittance, then develop the ellipse notation and show its usefulness in the transformation of beams' shapes in beam control problems. Building on the notation, we will define beam matching and how it is accomplished in periodic transport lines. Finally, we will give some emittance measurement techniques. These techniques will be divided into destructive and nondestructive methods. In a destructive method the beam propagation terminates in the measurement gear. Aside from the obvious disadvantage of destroying the beam for downstream users, destructive techniques require a finite amount of real estate and may not survive the high power density of present high-current, low-emittance accelerators. Reference 1 provides an excellent review of the emittance concept.

TRANSVERSE EMITTANCE

A beam can be considered an ensemble of particles, each with its own particular coordinates in terms of $(x, p_x, y, p_y, W, \phi)$. We will use a right-handed coordinate system with the z-direction along the beam axis, the y-direction in the vertical direction, and the x-direction in the horizontal plane. The p_x and p_y are the x- and y-components of the momentum in the x- and y-directions, respectively. The W and ϕ are differences in the particle energy and phase, respectively, from those of the synchronous particle. In characterization of the beam, one rarely measures p_x and p_y; rather, one measures the ratio of velocities (direction cosines) in the x-z and y-z planes. These ratios are referred to as horizontal and vertical divergences x' and y'. The values of x' and y' are given by

$$x' = v_x/v_z = p_x/p_z; \quad y' = v_y/v_z = p_y/p_z \ , \tag{1}$$

where v_x and v_y are the horizontal and vertical components of the velocity. The transverse emittance ε is the measure of the transverse phase space occupied by the beam:

$$\varepsilon = \iiiint dx \ dx' \ dy \ dy'/\pi \ . \tag{2}$$

The longitudinal emittance is a similar integral over W and ϕ.

In practice, one often uses 2D projections of the above 4D transverse emittance, for example,

$$\varepsilon_y = \int dy \; dy'/\pi = A_y/\pi \tag{3}$$

and

$$\varepsilon_x = \int dx \; dx'/\pi = A_x/\pi \;. \tag{4}$$

These equations define the vertical and transverse emittances, where A is the phase space area. In situations that involve acceleration, the emittance quoted is the normalized emittance given by

$$\varepsilon_y = \beta\gamma \; A_y/\pi \;, \tag{5}$$

where β and γ are the usually relativistic parameters given by

$$\gamma = E/mc^2 \;\;;\; \gamma = \left(1 - \beta^2\right)^{-1/2} \;. \tag{6}$$

With acceleration, the laboratory emittance will damp by $1/(\beta\gamma)$ as a result of the increase in p_z [Eqs. (1)]. Normally, the 6D emittance is conserved by Louisville's theorem, but if no energy is transfered to the average horizontal and vertical momentum of the beam, the normalized transverse emittances are conserved.

The convention is to quote the beam emittance for the portion of the beam inside the emittance, that is, ε_{90} or ε_{39} for the emittance containing 90% and 39% of the beam, respectively. As we will show, the 39% value corresponds to the root-mean-square (rms) emittance of a beam with a gaussian distribution. The units of emittance depend on the simulation code being used and are often π-radian-m or π-mradian-cm. The factor π is included to assure the reader that π has been factored out of the phase space area.

The concept of emittance evolved with the increasing understanding of accelerator dynamics. With the advent of computers that could analyze the large amount of data, the measurement of emittance became practical. The importance of emittance increased as high-intensity accelerators were developed with the requirement of low beam loss. We can locate and subsequently correct these beam losses by transporting the emittance down a model of the transport line and noting where the emittance envelope nears the apertures of the structure. Finally, understanding and minimizing the emittance is often important in accelerator applications.

PERIODIC FOCUSING AND THE ELLIPSE NOTATION

One of the consequences of phase-stable acceleration in radiofrequency (rf) accelerators is that one must operate with negative synchronous phase to get the longitudinal focusing required to contain the longitudinal emittance. This focusing introduces rf transverse defocusing. With the advent of quadrupoles, this rf defocusing and beams' envelopes could be controlled by the periodic insertion of magnetic quadrupoles as in drift-tube-linacs (DTLs) or by use of electric quadrupoles as in radiofrequency-quadrupoles (RFQs).

Figure 1 shows a simple periodic focusing system. The focusing properties are shown for the horizontal plane; F denotes horizontal focusing and D denotes defocusing. For quadrupoles, the focusing properties are the opposite in the vertical plane. Examination of a particle's path in the x, x' plane at periodic locations (homologous points) along this system—in the centers of the horizontally focusing quadrupoles in Fig. 1, for instance—would show that particles travel in elliptical paths. As a particle moves from one homologous point to another, the observed angle changes, a phenomenon called the phase advance. If one chooses different homologous points, the elliptical path will have a different shape. Often in a long periodic transport line, scraping and nonlinear effects can cause the nonuniform phase advance, leading the beam emittance to evolve into an elliptical shape. Typically, beams enter RFQs with very nonelliptical shapes but emerge with very elliptical shapes. Hence, for many simulations, beams are represented as ellipsoids in two, four, or six dimensions.

These ellipses can be traced through linear transport lines. The equation of an ellipse can be written as

$$\gamma x^2 + 2\alpha x x' + \beta x'^2 = A/\pi = \varepsilon , \tag{7}$$

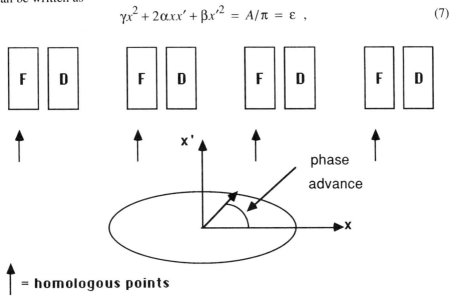

Fig. 1. Periodic transport line and elliptical path of particles at indicated homologous positions.

where

$$\beta\gamma - \alpha^2 = 1 \tag{8}$$

and α, β, and γ are the Twiss or the Courant-Snyder[2] parameters or ellipse parameters.

The major axis a, minor axis b, and orientation angle Θ of the ellipse can be expressed in terms of the Twiss parameters as

$$\tan 2\Theta = \frac{2\alpha}{(\gamma - \beta)}$$

$$\frac{a}{b} = \frac{\alpha}{\beta} + \frac{(\beta + \gamma)}{2} + \frac{\sqrt{(\beta + \gamma)^2 - 4}}{2} \ . \tag{9}$$

The Twiss parameters and the emittance as given in Eq. (7) have unique relationships with the ellipse (Fig. 2). In particular, the maximum values of x and x' are given by

$$x_{max} = \sqrt{\beta\varepsilon} \ ;$$

$$x'_{max} = \sqrt{\gamma\varepsilon} \ . \tag{10}$$

By using the definitions shown in Fig. 2, we can derive the values of emittance and Twiss parameters of a measured elliptical shape.

For nonuniform particle distributions, the Twiss parameters can be expressed in terms of the second moments M integrated over the particle distribution:

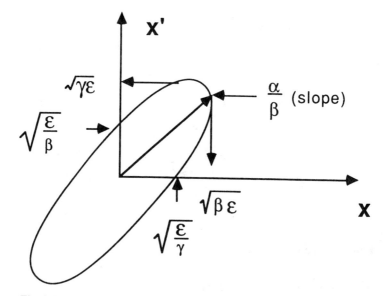

Fig. 2. Ellipse characteristics related to Twiss parameters and emittance.

$$M_{xx} = \iint x^2 \; \rho(x,x') \, dx \, dx' / \text{NORM} \, ,$$

$$M_{x'x'} = \iint x'^2 \; \rho(x,x') \, dx \, dx' / \text{NORM} \, ,$$

$$M_{xx'} = \iint xx' \; \rho(x,x') \, dx \, dx' / \text{NORM} \, ,$$

$$\text{NORM} = \iint \rho(x,x) \, dx \, dx' \; ;$$

(11)

where $\rho(x,x')$ is the particle density. Defining the x' limits of integration in terms of x from Eq. (7) and the x' limits of the integrals in Eqs. (11) in terms of the Twiss parameters in Eqs. (10), we can solve for the Twiss parameters in terms of the moments:

$$\beta = \frac{M_{xx}}{\sqrt{M_{xx}M_{x'x'} - M_{xx'}^2}}$$

$$\gamma = \frac{M_{x'x'}}{\sqrt{M_{xx}M_{x'x'} - M_{xx'}^2}}$$

$$\alpha = \frac{-M_{xx'}}{\sqrt{M_{xx}M_{x'x'} - M_{xx'}^2}} \; .$$

(12)

The denominator in Eqs. (12) is referred to as the rms emittance:

$$\varepsilon_{\text{rms}} = \sqrt{M_{xx} \, M_{x'x'} - M_{xx'}^2} \; .$$

(13)

Equations (11) and (12) define the rms parameters of the beam. Lapostolle[3] and Sacherer[4] showed that beam dynamics were a function of the beam rms values. If we transmit various distributions, all having the same rms values, through a transport line, then the various output beams would all have the same rms values. For first-order optics, the shape or distribution is immaterial. Therefore, if we want to simulate the behavior of a beam in a transport line, we can take an equivalent distribution with the same rms values, run it through our simulations, and predict the output rms values. Uniform ellipses are often chosen because they are relatively straightforward to propagate through simulations.

TRANSFORMATION OF ELLIPSES

Most beam simulation codes use a transfer matrix concept; that is,

$$\begin{bmatrix} x \\ x' \end{bmatrix}_f = [R] \begin{bmatrix} x \\ x' \end{bmatrix}_i \, ,$$

(14)

where R is the transfer matrix and i and f denote initial and final values, respectively. Examples of R-matrices are

$$R_{\text{drift}(L)} = \begin{bmatrix} 1 & L \\ 0 & 1 \end{bmatrix} \tag{15}$$

and

$$R_{\text{lens}} = \begin{bmatrix} 1 & 0 \\ -1/f & 1 \end{bmatrix} . \tag{16}$$

A simple drift of a particle with x- and x'-coordinates is

$$\begin{bmatrix} x \\ x' \end{bmatrix}_f = \begin{bmatrix} 1 & L \\ 0 & 1 \end{bmatrix} \begin{bmatrix} x \\ x' \end{bmatrix}_i = \begin{bmatrix} x + Lx' \\ x' \end{bmatrix} . \tag{17}$$

The R-matrix for a series of beam-transport elements is simply the matrix product of the series of matrices representing the elements. For example, the R-matrix for a thin-lens problem is

$$R = \begin{bmatrix} 1 & p \\ 0 & 1 \end{bmatrix} \begin{bmatrix} 1 & 0 \\ -1/f & 1 \end{bmatrix} \begin{bmatrix} 1 & q \\ 0 & 1 \end{bmatrix} , \tag{18}$$

where the three matrices represent (1) the drift q from the object to the lens, (2) the lens, and finally (3) the drift p from the lens to the image. In general, the beam has six coordinates, and hence, a six-dimensional matrix is required. For many cases, the matrix is block diagonal with 2 x 2 matrices along the diagonal when the various planes are uncorrelated. In cases that include dipole elements, skew quadrupoles, or rf deflectors, the 2D planes are correlated and the 6D matrix will have off-diagonal elements. The measurement of correlations is possible but difficult. In practice, most emittance studies are based on 2D measurements.

The ellipse equation, Eq. (7), can be written in matrix notation as

$$\begin{bmatrix} x \\ x' \end{bmatrix}^T \begin{bmatrix} \sigma \end{bmatrix}^{-1} \begin{bmatrix} x \\ x' \end{bmatrix} = 1 = \begin{bmatrix} x \\ x' \end{bmatrix}_f^T \begin{bmatrix} \sigma \end{bmatrix}_f^{-1} \begin{bmatrix} x \\ x' \end{bmatrix}_f = \begin{bmatrix} x \\ x' \end{bmatrix}^T \left[R\sigma R^T \right]^{-1} \begin{bmatrix} x \\ x' \end{bmatrix} , \tag{19}$$

where

$$\begin{bmatrix} \sigma \end{bmatrix} = \varepsilon \begin{bmatrix} \beta & -\alpha \\ -\alpha & \gamma \end{bmatrix} . \tag{20}$$

From Eqs. (19) and (20), we see that ellipses can be transformed by means of the "sigma"[5] matrix

$$\begin{bmatrix} \sigma \end{bmatrix}_f = \begin{bmatrix} R \end{bmatrix} \begin{bmatrix} \sigma \end{bmatrix}_i \begin{bmatrix} R \end{bmatrix}^T . \tag{21}$$

Once we know the transfer matrix for a beam line, the initial emittance, and the Twiss parameters, we can find the final emittance shape or Twiss parameters by the matrix multiplication of Eq. (21).

So-called envelope codes use Eq. (21), where the envelope refers to the radius x of some fraction of the beam:

$$x^2 = \varepsilon\beta = \left[\sigma\right]_{11} . \tag{22}$$

Examining the beam envelopes is useful for designing beam lines when one wants to avoid scraping on apertures. The envelope code execution is usually very fast. Particle tracing codes use the relationship

$$\begin{bmatrix} x \\ x' \\ y \\ y' \\ W \\ \phi \end{bmatrix}_f = \left[R\right] \begin{bmatrix} x \\ x' \\ y \\ y' \\ W \\ \phi \end{bmatrix}_i , \tag{23}$$

where Monte Carlo methods are used to generate the distribution of particles having the emittance parameters of interest. The final Twiss parameters are obtained from a moments analysis [Eqs. (12)] on the final distribution.

In general, a beam can be described by a gaussian distribution whose beam fraction F is given by

$$F = 1 - e^{-\left(\frac{\varepsilon}{2\varepsilon_{rms}}\right)} . \tag{24}$$

If the beam is gaussian, the emittance of 39% of the beam is the value of the rms emittance. Allison[6] showed that the emittance of a fraction of a gaussian beam is given by

$$\frac{\varepsilon_{rms}(F)}{\varepsilon_{rms}} = 1 + (1 - F) \ln (1 - F)/F . \tag{25}$$

Extrapolating this expression to $F = 1$, we can evaluate the rms emittance of the total beam in the presence of background. Figure 3 from Ref. 6 is a plot of Eq. (25) using data from a Penning source. It shows the extrapolated rms value of this beam core. Note that greater than 80% of the beam is gaussian. One often sees beams of this character—beams that have a "halo" or, in some cases, appear to be the superposition of two beam shapes.

APPLICATIONS

We frequently wish to maximize the beam on a target or minimize the beam size xf at a target using a lens F followed by a drift L to the target. Using Eqs. (15) and (16) to describe the R-matrix, Eq. (21) to describe the transformation of the sigma matrix, and Eq. (22) to define xf, we find that

$$\sigma_{11}^f = \sigma_{11}^i \left(1 - \frac{L}{F}\right)^2 + \sigma_{12}^i \left(1 - \frac{L}{F}\right) + \sigma_{22}^i L^2 \ . \tag{26}$$

Differentiating Eq. (26) with respect to the focal length to minimize the size and then solving for the beam size, we find

$$\sigma_{11}^f = \frac{\varepsilon^2 L^2}{\sigma_{11}^f} \ ; \ x_f^2 = \frac{\varepsilon^2 L^2}{x_i^2} \ ; \ x_{min} = \frac{\varepsilon L}{x_{lens}} \ . \tag{27}$$

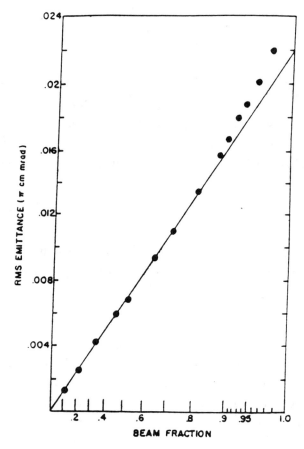

Fig. 3. Plot of ε_{rms} versus $1+(1-F)\ln(1-F)/F$ for Los Alamos Penning source; from Ref. 6.

Equations (27) show that to minimize the beam size on the target, we must minimize the emittance and length and maximize the beam size in the lens.

Beam transports are limited in the emittance size that they can accept without beam losses caused by scraping. The limit is the acceptance of the transport line. We will now explore this acceptance and how we can minimize transmission losses and emittance growth in a transport line. The R-matrix for one period of a periodic transport line can be written in terms of the Twiss parameters that describe the elliptical path of the particles when observed at successive homologous points:

$$R = \begin{bmatrix} \cos\mu + \alpha\sin\mu & \beta\sin\mu \\ -\gamma\sin\mu & \cos\mu - \alpha\sin\mu \end{bmatrix}, \qquad (28)$$

where μ is the phase advance in Fig. 1.

We can transport this ellipse to a circle by the following transformation:

$$\begin{bmatrix} \eta \\ \eta' \end{bmatrix} = \begin{bmatrix} \dfrac{1}{\sqrt{\beta}} & 0 \\ \dfrac{\alpha}{\sqrt{\beta}} & \sqrt{\beta} \end{bmatrix} \begin{bmatrix} x \\ x' \end{bmatrix}, \qquad (29)$$

where the Twiss parameters are those of Eq.(28). This reference frame is called the "normalized frame," in which particles move in circular orbits with μ phase advance as they move from one homologous point to the next. Beams that have the same Twiss parameters as Eq. (28) will have a circular shape that remains unchanged from one homologous point to the next because of the circular motion of the particles. Such a beam is defined as "matched" to the transport line. Unmatched beams will transform to ellipses that rotate around the origin as they move from one homologous point to the next (Fig. 4).

Equation (29) shows that the projection of the beam to the η-axis is equal to the x-projection after a scaling of $1/\sqrt{\beta}$. Hence, observation of the x- or y-projections of the beam at homolgous points on a beam line show that matched beams always have the same shape and width. On the other hand, unmatched-beam widths increase and decrease as the ellipse rotates from having its major axis along the η-axis to having its minor axis along the η- (or x-) axis. If a matched and an unmatched beam have the same emittances, then the major axis of the unmatched beam will be greater than the radius of the matched beam. Hence, the unmatched beam will be more susceptible to beam scraping and nonlinear effects in focusing elements.

Matching the beams to the structure allows the maximum emittance to be transmitted without scraping. Periodic apertures and the elliptical path of the particles as they move along a transport line combine to create an elliptical acceptance. The ellipse has the Twiss parameters of Eq. (28) and emittance determined by Eq. (22), where x is the

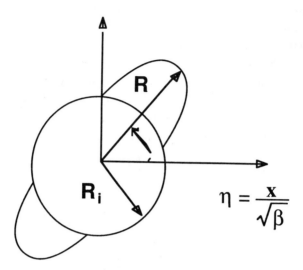

Fig. 4. Normalized frame showing shapes of a matched beam (circle) and unmatched beam (ellipse). All particles move in circular orbits in this frame.

radius of the aperture. Particles outside the acceptance of a transport line are lost. Often nonlinear effects cause the phase advance to change as particles move farther from the beam-line center. As a result, the beam can "wind up" inside a circle whose radius is the major axis of the beam ellipse. Such a beam can evolve to a matched beam with a growth in its apparent emittance.

The mismatch factor MM, as discussed in Ref. 7, is a measure of how well the beam is matched to a structure. It is the ratio of the difference between the major axis of the unmatched beam and the radius of the matched beam to the radius of the matched beam:

$$MM = \sqrt{1 + \frac{\Delta}{2} + \sqrt{\Delta + \frac{\Delta^2}{4}}} - 1 \ , \tag{30}$$

where

$$\Delta = \beta_b \gamma + \beta \gamma_b - (2\alpha\alpha_b - 2) \ . \tag{31}$$

Both beams have the same emittance area. The Twiss parameters above are those of Eq. (28) and those of the beam (b). The MM factor is an estimate of the betatron oscillation, scraping, nonlinear effects, and transmission of a beam through a structure. The potential for growth resulting from a mismatched beam filling the areas of the enclosing matched shape is

$$\frac{dA}{A} = (MM - 1)^2 - 1 \ . \tag{32}$$

The MM factor is the fractional area increase of the ellipse enclosing two superimposed ellipses of equal area.

Various methods are used to match a beam to a transport line. If upstream focusing elements are available, we can measure the emittance at the entrance of the line and then adjust the upstream elements to change the beam emittance to have the matched beam parameters. If no emittance measuring gear is available, we can measure profiles at homogeneous points and minimize the variations in the beam profiles widths by again adjusting the upstream elements. Alternatively, we can adjust the upstream elements to maximize the transmission. This method works when the beam emittance is approximately equal to or greater than the acceptance of the line. Another method used at the Los Alamos Meson Physics Facility (LAMPF) is the "quad unroll" technique. Small variations in quadrupoles fields will not change the matched Twiss parameters of the line but can greatly change the total phase advance through the line. Hence, an unmatched beam ellipse will rotate (Fig. 4) in the "normalized" frame, and the laboratory emittance and its x- or y-projections will alternately increase and decrease in size as described above.

EMITTANCE MEASUREMENT METHODS

Six methods of emittance measurements will be reviewed. The first four—slit-slit, slit and collector, electric sweep, and pepper pot—are destructive. The wire-shadow and tomography methods can be nondestructive.

The slit-and-collector method is a modification of the two-slit method as depicted in Fig. 5 from Ref. 1. Its objective is to measure the 2D emittance density distribution:

$$\rho_2 (x, x') = \int\int \rho_4 \; dy \; dy' \; . \tag{33}$$

For a horizontal emittance measurement, a thin, vertical, upstream slit selects a small portion of the emittance to be transmitted downstream. The slit position identifies the x, or horizontal, position of the beamlet, but all values of y and y' are allowed. In the slit-slit method, a second downstream vertical slit is scanned across the beamlet, and the transmitted beam is measured, typically with a Faraday cup, as a function of the second slit position. The slit positions, the transmitted current values, and the distance L separating the two slits are used to generate the angular (x') distribution of the beamlet. Repeating this process for different upstream slit positions produces an x, x' density distribution that is used in a moments analysis, where the integration is replaced by summation. For example,

$$M_{xx} = \frac{\Sigma_i \Sigma_j \; x_i^2 \rho (x_i, x_j')}{\Sigma_i \Sigma_j \; \rho (x_i x_j)} \; . \tag{34}$$

The method is repeated with horizontal slits, which move vertically, to obtain a measurement of the vertical emittance.

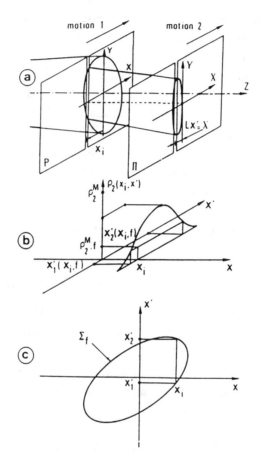

Fig. 5. The "two-slit method" from Ref. 1. (a) Schematic representation of the principle, (b) profile of projected density for a chosen position x_i of the upstream slit, (c) the range of x' of a particular equidensity contour E_f corresponding to the fraction f of the maximum density ρ_2^M in Fig. 5a.

The process can be expedited if the second slit is replaced with a sandwich of vertical detectors as depicted in Fig. 6. The detectors are typically secondary emission monitors with low electric fields applied at the surface to sweep the liberated secondary electrons away from the detector. We can produce the electric field by positively biasing a foil or grid just upstream of the collector or negatively biasing the collectors themselves. Without the bias, many electrons will be recollected on adjacent monitors with a corresponding distortion of the apparent angular distribution. Using the detector positions, their individual signals and the distance L between the collector and slit, we can again determine the angular distribution. Repeating this process can produce an x,x' density distribution. Figure 7 shows the slit-and-collector equipment used by the Accelerator

Technology Division at Los Alamos National Laboratory. Thin strips of copper serve
as secondary emission detectors. This design is based on equipment used earlier at
LAMPF, Brookhaven National Laboratory, and the Fermi National Accelerator
Laboratory.[8] Figures 8 through 10 show a typical isometric of the beam emittance
distribution, a contour plot of the distribution, and the results of a moments analysis
of the beam for various thresholds and percentages of total beam.

The slit-slit and the slit-collector methods both have advantages and disadvantages
over each other. The slit-collector method takes less time because no motion is required
to measure the angular distribution for each slit position. The slit-slit method has only
one particle detector and does not require calibration of the many detectors used in the
slit-collector method. With the slit-slit method, the angular step size can be varied easily
with a simple adjustment of the step size of the final slit. However, we can also obtain
this advantage with the slit-collector method by moving the collector downstream to
provide sufficient resolution and by adjusting the slit-to-collector step-size ratio to span
the total range of angles. The slit-slit and slit-collector distances must often be tailored
to the beam emittance to make optimal use of the actuator travel range and angular resolution. The emittance is time averaged because the total emittance is determined from
many different beam pulses. For low-energy, positive beams, the presence of the slit can
alter the apparent emittance shape. This effect is caused by the vast amount of secondary
electrons, which are produced by the beam striking the slit, changing the neutralization
of the upstream beam and changing the amount of space-charge defocusing.

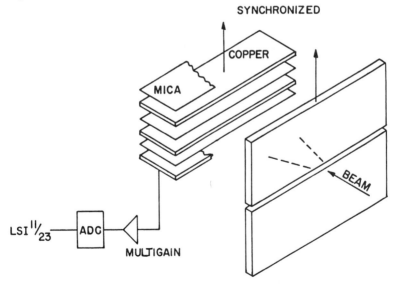

Fig. 6. Schematic representation of the "slit and collector" method.

Fig. 7. Photograph of a slit and collector assembly used at Los Alamos National Laboratory.

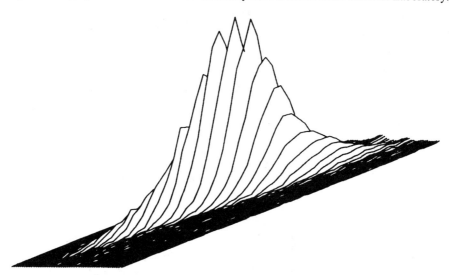

Fig. 8. Isometric display of the transverse emittance of a RFQ
mounted on the Accelerator Test Stand at Los Alamos.

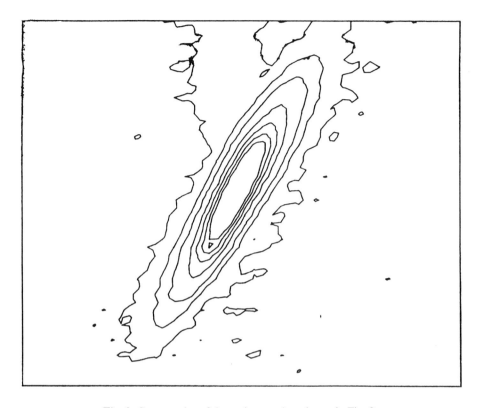

Fig. 9. Contour plot of the emittance data shown in Fig. 8.

Both methods are very robust; currents densities of 80 to 100 mA over a 4- to 6-mm diameter have not significantly damaged upstream slits made of graphite. At Los Alamos, pyrolytic graphite has been used successfully with 2- and 5-MeV beams having a 5-Hz repetition rate and a pulse width of 50 to 100 μs. The equally robust copper-mica sandwich collectors have survived exposure to the total beam. Both slit and collectors are water cooled.

In emittance measurements, two sources of errors are space charge and the finite slit and collector sizes. The ribbon beam that passes through the slit can expand as a result of space charge effects[7] and artificially increase the observed distribution. This effect is insignificant if

$$L < \sqrt{\frac{4}{kJ}} \ , \qquad (35)$$

where J is the current density and

$$k = \frac{e}{mv^3 \varepsilon_0} \ . \qquad (36)$$

Reanalyze — Version 2.7

Run no. 259   ES 3 15-OCT-87 11:14:38 Vertview egy = 38.0   Uex= 0.00   Vsu= 0.00 Xe= 0.00 fc2= 102.00
Pick-up= 0.00 Bunch cav= 0.00   Bunch phase= 0.00 RFQ= 700.00 hee1= 0.00   hee2= 0.00 labt% 39
s1= −5/1.01   s2= .4/−1.3   DTL cav=   0.00 DTL phase=   0.00 energy= 2.0700 step size= 0.085
slit off= 0.0000   coll off= 0.1600   comment: MMF=0.06

Moments analysis based on density with no adjustments

Maximum counts =   2675.   Total counts =   1 72751.   Beam current =   0.0

thresh	Xave	X'ave	beta	gamma	alpha	Erms	Etot	theta	c**2 % abov	% thru
0.01–1.457E–02	3.11	1.295E–02	1.21	−0.749	1.07	16.9	89.6	121.	105.	106.
1.00–1.553E–02	3.52	1.509E–02	177.	−1.30	0.637	5.79	89.6	177.	91.4	100.
2.00–1.515E–02	3.42	1.669E–02	197.	−1.51	0.528	4.16	89.6	197.	84.3	96.6
4.00–1.478E–02	2.74	2.340E–02	249.	−2.20	0.340	2.63	89.5	249.	74.5	90.0
6.00–1.484E–02	2.73	2.367E–02	257.	−2.26	0.302	2.15	89.5	257.	67.5	86.5
8.00–1.458E–02	2.67	2.333E–02	260.	−2.25	0.277	1.86	89.5	260.	61.6	83.6
10.00–1.342E–02	2.77	2.315E–02	262.	−2.25	0.251	1.59	89.5	262.	56.6	80.1
12.00–1.554E–02	2.58	2.296E–02	261.	−2.24	0.232	1.42	89.5	261.	52.2	77.3
20.00–1.442E–02	2.66	2.240E–02	272.	−2.26	0.176	0.963	89.5	272.	38.4	66.9
30.00–1.675E–02	2.30	2.123E–02	262.	−2.14	0.136	0.686	89.5	262.	26.8	57.2
40.00–1.937E–02	2.18	2.236E–02	295.	−2.37	9.892E–02	0.465	89.5	295.	18.5	46.0
50.00–1.810E–02	2.19	2.136E–02	277.	−2.22	8.243E–02	0.376	89.5	277.	12.3	40.1
60.00–1.955E–02	2.16	2.443E–02	330.	−2.66	5.814E–02	0.255	89.5	330.	7.54	30.1
70.00–1.979E–02	2.14	2.123E–02	283.	−2.24	4.147E–02	0.177	89.5	283.	4.34	22.7
80.00–2.341E–02	1.90	2.159E–02	360.	−2.56	3.138E–02	0.133	89.6	350.	2.08	17.8
90.00–2.540E–02	1.58	3.657E–02	452.	−3.94	1.556E–02	6.640E–02	89.5	452.	0.543	9.38

Beam fractions: 39.4%, 63.2%, 77.7%, 86.5%, 91.8%, 95.0%, (1 thru 6 sigma)

thresh	Xave	X'ave	beta	gamma	alpha	Erms	Etot	theta	c**2 % abov	% thru
50.69–1.928E–02	2.14	2.160E–02	290.	−2.29	8.007E–02	0.365	89.5	290.	11.9	39.3
23.83–1.726E–02	2.40	2.259E–02	267.	−2.24	0.156	0.819	89.5	267.	33.5	62.3
11.73–1.552E–02	2.59	2.328E–02	263.	−2.26	0.235	1.44	89.5	263.	52.8	77.7
6.02–1.447E–02	2.72	2.361E–02	258.	−2.26	0.300	2.14	89.5	258.	67.4	86.4
3.47–1.466E–02	2.89	2.169E–02	236.	−2.03	0.375	2.93	89.5	236.	76.7	91.7
2.49–1.503E–02	3.31	1.733E–02	203.	−1.59	0.491	3.71	89.5	203.	81.5	95.1

Extrapolated rms emittance assuming gaussian beam

thru % beam		rms emittance	normalized (2.00 MeV)
77.7	0.394	+/− 1.724E–02	2.573E–02
86.5	0.414	+/− 1.960E–02	2.703E–02
91.8	0.439	+/− 2.912E–02	2.868E–02
95.0	0.483	+/− 5.649E–02	3.153E–02

Fig. 10. Results of a moments analysis on the data of Fig. 8. The Twiss parameters, the emittance, the rms emittance, and beam fraction are displayed versus fraction of the peak (ρ_2^M) of the emittance distribution.

The effects[9] of finite slit and collector sizes depend on the area and shape of the emittance to be measured. If $2s$ is the slit width and $2c$ is the collector width, then the measured emittance values are given by

$$\varepsilon = \varepsilon_t \sqrt{1 + \frac{\Delta}{\varepsilon_t}} \; ;$$

$$\beta = \frac{\left(\beta_t + \dfrac{s^2}{3\varepsilon_t}\right)}{\sqrt{\left(1 + \dfrac{\Delta}{\varepsilon_t}\right)}} \; ; \qquad \alpha = \frac{\left(\alpha_t + \dfrac{s^2}{6L\varepsilon_t}\right)}{\sqrt{1 + \varepsilon_t}} \; ; \tag{37}$$

$$\gamma = \frac{\left(\gamma_t + \dfrac{s^2}{6L\varepsilon_t}\right)\left(1 + \dfrac{c^2}{s^2}\right)}{\sqrt{1 + \dfrac{\Delta}{\varepsilon_t}}} \; .$$

In Eq. (37), (t) denotes true values and

$$\Delta = \frac{s^2}{3}\left(\frac{\beta_t}{L}\left(1 + \frac{c^2}{s^2}\right) + \gamma_t - \frac{\alpha_t}{L}\right) . \tag{38}$$

Conversely, the true values can be found from the measured values of Eqs. (37) and (38). The required number of position and angle samples required for accurate reconstruction of the emittance has been determined empirically.[6] More than five samples in angle and position were required across the center region of the emittance.

Many detectors use secondary emission of electrons from the detector surface. Salehi and Flinn[10] have measured the secondary emission yield Y for electrons to be

$$\frac{Y}{Y_m} = \frac{\dfrac{2E}{E_m}}{1 + \left(\dfrac{E}{E_m}\right)^{3.7\frac{Z}{A}}} , \tag{39}$$

where A and Z are the atomic mass and charge, respectively, of the surface material and

$$Y_m \approx 1.3 \; @ \; E_m = 475 \text{ keV} . \tag{40}$$

The measured yield for protons[11] is given by

$$Y = BE^{-0.7} \csc\theta , \tag{41}$$

where θ is the incident angle and the constant B varies between 1.1 and 2.2.

Figure 11 shows the beam-energy dependence of the effective negative charge flowing off a copper strip that is biased to repel any liberated secondary electrons. This plot[12] shows that for energies less than 2 MeV, H⁻ and H⁰ beams produce more signal than protons. Above 2 MeV, however, the secondary emission becomes sufficiently small that these beams would produce more signal if they were stripped to protons.

A clever variation of the two-slit method is the so-call "electric sweep" method.[13] Figure 12 shows the basic idea. The ramped electric field moves different angles of the beam across the slit; then this selected beam is collected in a Faraday cup. By digitizing both this Faraday cup current versus time and the electric field versus time, we can determine the beam angular distribution. This procedure is repeated as the whole device is stepped in position across the beam; the resulting data is a density matrix in emittance space. This method is as fast as the slit-and-collector method because one can measure the angular distribution in one macropulse. In addition, one can change the angular range by simply changing the ramp voltage. This method typically gives finer angular resolution than the slit and collector with smaller values of L. The disadvantages are that measuring beams with energies above 1 MeV becomes a difficult high-voltage problem and that 10 to 100 μs are required to measure the angular distribution. This time restriction limits the bandwidth of observable time-dependent emittance changes.

Figure 13 is a composite photograph of a small electric sweep scanner having a total length of 2 in. A contour plot of a 100-keV injector beam is shown in Fig. 14. The aberrations from the accelerating column are evident in this figure. After being accelerated in a RFQ, this beam looks very elliptical.

The so-called "pepper pot" method uses an array of identical holes (pinholes) in a plate oriented in the x-y plane. This plate is placed in the beam upstream of a beam-sensitive plate made of material such as gadolinium oxysulfide, known commercially as RaRex™. This material contains a wavelength shifter to produce light in the visible range of a camera. These plates can have an energy conversion efficiency of 15%.

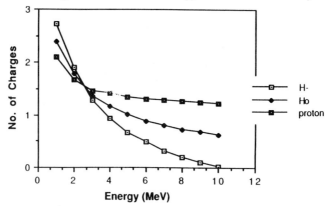

Fig. 11. Effective negative charge leaving a copper strip versus incident beam energy. The strip is assumed to be electrically biased to prevent the recollection of secondary electrons.

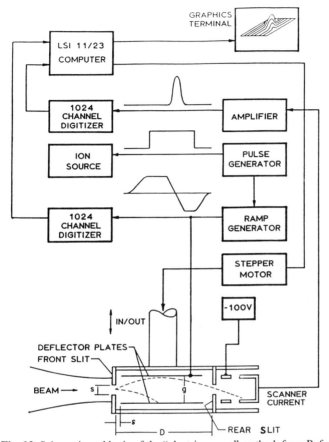

Fig. 12. Schematic and logic of the "electric sweep" method; from Ref. 6.

The transmitted beamlets strike the plate and produce an image that is proportional to $\rho_4(x,x',y,y')$. Figure 15 shows a schematic of the method. The width of the spot is a measure of the angular divergence of the beam; the detailed intensity distribution gives the angular distribution as in the slit and collector method. The advantage of this method is that we can get a complete 4D emittance measurement in one macropulse, although the data reduction is very time consuming.

Often only the centroids of the beamlets are examined. The beam focusing can be determined by the magnification of the separation of the pinhole images. Aberrations can be evaluated by making polynomial fits to the image locations. We can determine and correct chromatic aberrations by measuring the image width or angular divergence versus position. Figure 16 shows two images for different pinhole positions. The change with width for an off-axis beam is very apparent. By reducing the energy spread of the beam, the chromatic aberrations can be reduced. Figure 17 shows how pinhole images were used to determine the effects of using an rf buncher to reduce the chromatic effects.

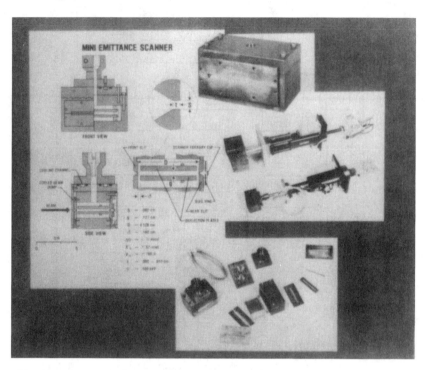

Fig. 13. Photograph of the "electric sweep" scanner, its parts and design drawing.

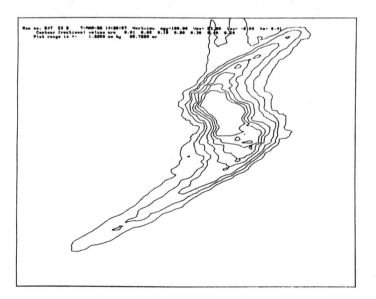

Fig. 14. Contour plot of the emittance of a 100-keV beam as measured by an "electric sweep" scanner.

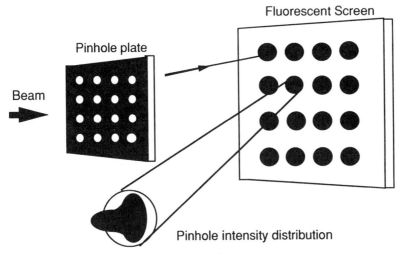

Fig. 15. Schematic of the "pepper pot" method.

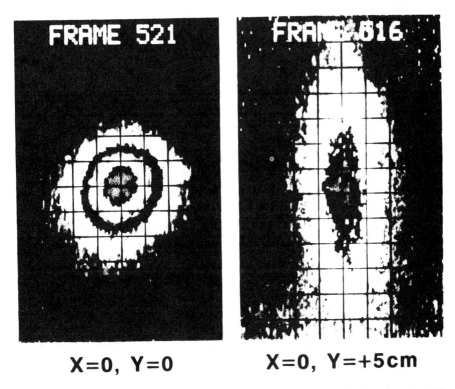

Fig. 16. Images of two pinholes at difference distances from the beam axis. Note that the width increases with increasing distance. A symmetric increase in width as the pinhole distance from the beam axis increases is indicative of chromatic aberrations.

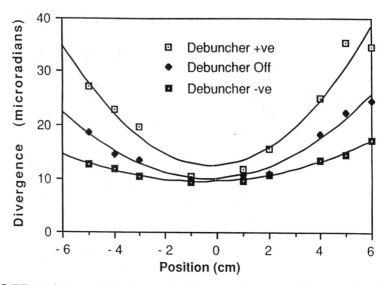

Fig. 17. Effects of using an rf buncher to either increase or decrease the chromatic aberrations.

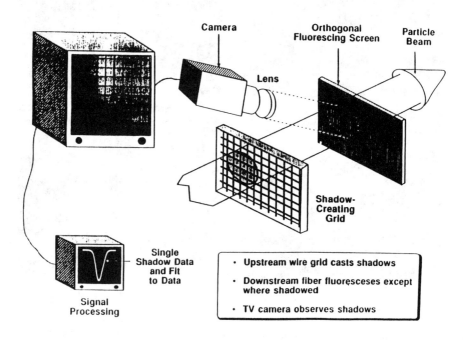

- Upstream wire grid casts shadows
- Downstream fiber fluoresceses except where shadowed
- TV camera observes shadows

Fig. 18. Schematic of the "wire shadow" method of measuring beam characteristics.

A nearly nondestructive method for measuring the transverse emittance is the "wire shadow" method (Fig. 18) being developed at Los Alamos. An upstream shadow grid is placed in the beam, and the resulting beam shadows are observed on a downstream grid of fluorescing wires. The shadows are the inverse of the light signature that we would observe if the wire were replaced by a slit. The shadow patterns are recorded and digitized by a video camera. With this device, we can not only measure the usual x,x' and y,y' 2D transverse emittance but also the x-y, the x-y', and the x'-y correlations. Only the absence of the x'-y' correlation prevents measurement of the $\rho_4(x,x',y,y')$ density distribution. With sufficiently small wires, typically 25μm, only a small amount of beam is scattered; hence, this method can be justly called nondestructive. If the shadows are ignored, the overall light distribution on a wire is a direct measurement of the horizontal or vertical beam profile. By monitoring just one shadow, one can quickly tune upstream beam elements to minimize the beam divergence by minimizing the width of the shadow. This tuning also increases the depth of the shadow as shown in Fig. 19. Just as with the pepper pot measurement, we can use the shadow centroids to determine the focusing and aberrations of the beam. The major concerns of this method are resolution and survival of the fluor, survival of the wire, and shadow broadening caused by mechanical vibration.

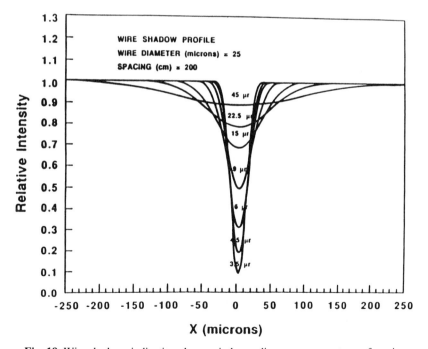

Fig. 19. Wire shadows indicating changes in beam divergence as upstream focusing elements are adjusted.

The last method to be discussed is based on tomography. The emittance is reconstructed from position profiles, which are projections of the emittance onto the position axis. The profiles can be easily measured, either by use of wire scanners or by measurement of the beam excitation of the residual gas[14] in the beam line. The beam excitation method is completely nondestructive. The minimum number of profiles is three. One can gain these profiles either by using a single upstream lens varied to give different focusing and therefore different profiles or by measuring profiles at three different positions. In the latter case, the simplest situation[15] is three locations separated by drift spaces with the center position preferably at a beam waist.

Figure 20 schematically shows the rationale for choosing the optimal viewing positions. Viewing position #1 is at a waist, so the beam emittance is an upright ellipse. This ellipse is transformed to the normalized frame $(\eta - \eta')$. With three views possible, the views should be separated by $120°$ as shown in Fig. 20. With viewing position #1 already chosen, the other views are at upstream and downstream drift distances from the waist position. We solve for drift L using Eqs. (20) through (22) and the fact that the point A in Fig. 20 will be a position extremum:

$$\begin{bmatrix} 1 & L \\ 0 & 1 \end{bmatrix} \begin{bmatrix} \beta & 0 \\ 0 & \gamma \end{bmatrix} \begin{bmatrix} 1 & 0 \\ L & 1 \end{bmatrix} = \begin{bmatrix} \sigma \end{bmatrix}_L \; ; \tag{42}$$

$$X_L = \sqrt{\beta_L \varepsilon} = \sqrt{\left(\beta + \frac{L^2}{\beta}\right)\varepsilon} \; ; \tag{43}$$

$$\begin{bmatrix} x \\ x' \end{bmatrix}_L = \begin{bmatrix} 1 & L \\ 0 & 1 \end{bmatrix} \begin{bmatrix} x \\ x' \end{bmatrix}_A \; ; \tag{44}$$

where

$$\begin{bmatrix} x \\ x' \end{bmatrix}_A = \begin{bmatrix} \sqrt{\beta} & 0 \\ 0 & \frac{1}{\sqrt{\beta}} \end{bmatrix} \begin{bmatrix} \eta \\ \eta' \end{bmatrix}_A = \begin{bmatrix} \sqrt{\beta} & 0 \\ 0 & \frac{1}{\sqrt{\beta}} \end{bmatrix} \begin{bmatrix} \frac{\sqrt{\varepsilon}}{2} \\ \frac{2\sqrt{\varepsilon}}{\sqrt{3}} \end{bmatrix} \; . \tag{45}$$

Using Eqs. (43) through (45), we can show that

$$L = \beta\sqrt{3} \; . \tag{46}$$

This exercise demonstrates that the optimal viewing distances depend on emittance shape. The method requires knowledge of the transfer matrices between the three or more views. Computer codes, TRACE[16] and TRANSPORT,[5] for example, are available to compute the transfer matrices. In the presence of space charge, the transfer matrices depend on the emittance. Therefore, one typically solves for the emittance and Twiss parameters in an iterative manner: one can solve for the emittance without space charge, input this emittance into the determination of the transfer matrix, then resolve

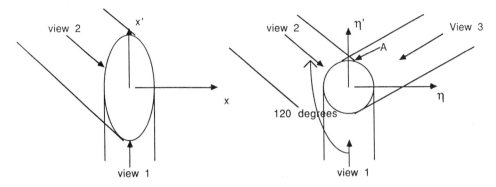

Fig. 20. Optimal viewing angles for tomographical reconstruction.

for the emittance using these transfer matrices. By using profile measurements other than along the x- and y-axes—along a line 45° to the x- and y-axes, for instance— we can measure the correlations and determine the full $\rho_4(x,x',y,y')$ density distribution.[17] The tomography method, which used the algorithm MENT[18] (Maximum ENTropy), has been compared with the slit-and-collector method using the 100-MeV LAMPF beam.[19] Figure 21 shows the contour plots from the two methods. Emittance areas are compared in Fig. 22. The agreement is good, and the mismatch factor between the results of the two methods was less than 0.15.

We also used MENT to reconsruct the transverse emittance[20] of the FMIT[21] accelerator. In this case, the intense beam precluded the use of wire scanners; the profiles were obtained by measurement of the visible light emitted following the beam interaction with the residual gas.

If we assume that the emittance density contours are concentric and have the same Twiss parameters, then the reconstruction of the emittance from profiles becomes simpler. By using Eq. (7) and methods similar to those used to derive Eqs. (12), we can show that the edges of the profiles, measured at various places in a beam transport, can all be transformed to one common place in the beam line and the various x and x' combinations will form the outer contour of the emittance distribution. In addition, if we transform the $\pm x$ values, which enclose equal areas of each profile, the corresponding x and x' combinations will all fall on an inner contour of the emittance distribution. Such reconstruction methods have been used successfully on accelerator beams whose shapes are nearly elliptical.

In conclusion, we have presented the emittance concept; we have presented how and why we often manipulate the emittance; and finally we have presented the basic ideas behind the many techniques for measuring the emittance.

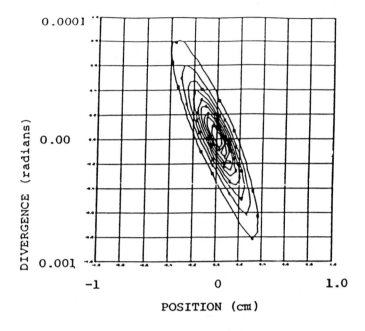

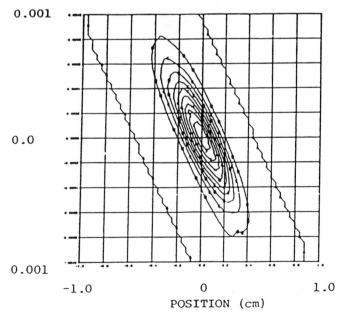

Fig. 21. Emittance contour plots from the "slit and collector" method and from the tomography method.

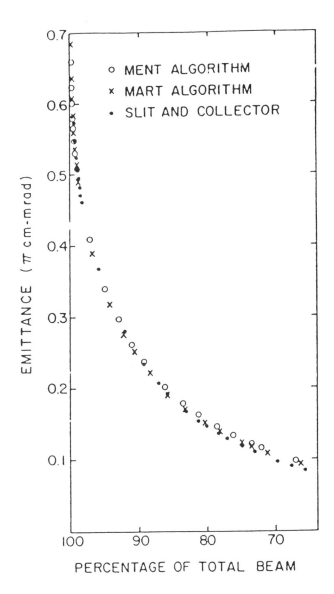

Fig. 22. Plots of emittance areas versus percentage of beam peak for the "slit and collector" method and for the tomography method.

REFERENCES

1. C. Lejeune and A. Aubert, "Emittance and Brightness Definitions and Measurement," Adv. in Electronics and Electron Physics, Supplement 13A, Applied Charged Particles Optics, A. Septier, ed., Part A, Academic Press, 129 (1980).

2. E. D Courant, H. S. Synder, "Theory of the Alternating Gradient Synchrotron," *Ann. Phys.* **3**, 1–48 (1958).

3. P. Lapostolle, "Possible Emittance Increase through Filamentation due to Space Charge in Continuous Beams," *IEEE Trans. Nucl. Sci.* **NS-18** (3), 1101 (1971)

4. F. J. Sacherer, "RMS Envelope Equations with Space Charge," *IEEE Trans. Nucl. Sci.* **NS-18** (3), 1105 (1971).

5. K. L. Brown, B. K. Lear, and S. K Howry, "TRANSPORT.360," Stanford Linear Accelerator Center Report SLAC-91 (1977).

6. P. W. Allison, "Some Comments on Emittance of H⁻ Ion Beams," 4th International Symposium on the Production and Neutralization of Negative Ions and Beams, Upton, New York (October 1986).

7. J. Guyard, M Weiss, "Use of Beam Emittance Measurements in Matching Problems," Proc. 1976 Linear Accel. Conf., Atomic Energy of Canada, AECL-5677 (1976), p. 254.

8. R. W. Goodwin, "Beam Diagnostics for the Fermi National Laboratory 200-MeV Linac," Proc. 1970 Proton Linear Accel. Conf., Batavia, Illinois, FNAL report LCO-058, 107 (1970).

9. R. L. Gluckstern and K. F. Johnson, private communication (March 1988).

10. M. Salehi, E. A. Flinn, *J. Phys. D: Appl. Phys.* **13**, 281 (1980).

11. A. Koyama, E. Yagi, H Sakairi, *J. Appl, Phys.* **15**, 1811 (1976).

12. R. Connolly, private communication, (December 1988).

13. P. W. Allison, J. D. Sherman, D. B. Holtkamp "An Emittance Scanner for Intense Low-Energy Ion Beams," *IEEE Trans. Nucl. Sci.* **NS-30** (4), 2204 (1983).

14. J. S. Frazer, "Developments in non-Destructive Beam Diagnostics," *IEEE Trans. Nucl. Sci.* **NS-28** (3), 2137 (1981).

15. C. Metzger, "Measurement of Beam Emittances and Beam Centering in the 800 MeV   Measurement Line of the PS Booster," CERN/SI/Int DL/69-10 (October 21, 1969).

16. K. R. Crandall, "TRACE—An Interactive Beam Transport Program," Los Alamos National Laboratory Report LA-5532 (October 1973).

17. G. N. Minerbo, O. R. Sander, R. A. Jameson, "Four-Dimensional Beam Tomography," *IEEE Trans. Nucl. Sci.* **NS-28** (3), 2231 (1981).

18. G. N. Minerbo, "MENT, a Maximum Entropy Algorithm for Reconstructing a Source from Projection Data," *Computer Graphics and Image Processing* **10**, 48 (1979).

19. O. R. Sander, G. N. Minerbo, R. A. Jameson, D. D. Chamberlin, "Beam Tomography in Two and Four Dimensions," Proc. 1979 Linear Accelerator Conference, Montauk, New York, Sept. 1979, Brookhaven National Laboratory report 51134, p. 314 (1979).

20. D. D Chamberlin, G. N Minerbo, L. E. Teel, J. D. Gilpatrick, "Noninterceptive Transverse Beam Diagnostics," *IEEE Trans. Nucl. Sci.* **NS-28** (3), 2347 (1981).

21. D. D. Armstrong, "The FMIT Accelerator," *IEEE Trans. Nucl. Sci.* **NS-30** (4), 2965 (1983).

Loss Monitors

Marvin Johnson
Fermi National Accelerator Laboratory*
PO Box 500
Batavia, IL. 60510

Introduction

This lecture will discuss the different types of detectors that are used as loss monitors, the physics processes that occur in these detectors, some comments on the electronics and some practical hints for building them. Since loss monitors are very similar to detectors used in experimental high energy physics, many of the references will be from high energy physics instrumentation.

When particles pass through material they can interact in a number of ways. For loss monitors only the electromagnetic interaction is important. All other processes are smaller by several orders of magnitude. Electromagnetic interactions either excite or ionize the material. Excitation based devices are scintillators and are discussed in section I. Ionization devices are usually some type of gas device with the most common one being the ion chamber. These are discussed in section II. Section III gives a brief discussion of solid state loss monitors while section IV describes the use of colliding beam experimental detectors themselves as loss monitors.

I. Scintillator Loss Monitors

What is scintillation? [1,2,3] When a particle passes through material it can raise the molecules to an excited state. Scintillation is light from the de-excitation of these molecules. There are two broad classes of scintillator which produce the light by different mechanisms.

The first type consists of inorganic crystals doped with a metal such as chromium or thallium[4,5]. Typical concentrations of thallium are around 10^{-3} mole basis. An impurity atom replaces one of the atoms in the crystal much like an impurity in a semiconductor. The impurities are all donor atoms so they are ionized in the crystal lattice. The excited states of the impurities are in

* Operated by Universities Research Association, Inc. under contract with the U. S. Department of Energy.

the UV or blue part of the visible spectrum. The energy levels of these states are shifted by the field of the crystal into the visible spectrum.

Since the impurity concentration is so low, almost all of the energy loss from a particle passing through the material goes into creating electron-hole pairs in the bulk crystal. These electron-hole pairs diffuse through the crystal to impurity sites where the electron combines with the ionized impurity to neutralize it. A hole then combines with the neutralized atom giving an ion back and an emitted photon. Thus the impurity centers serve as sites for electron-hole recombination. There are a large number of these types of scintillators. Sodium iodide doped with thallium is probably the most commonly used one. Another example is aluminum oxide doped with chromium. This type of scintillator is not used very often for loss monitors but they are sometimes used for fluorescent screens to measure beam sizes[6].

The second type of scintillator is an organic molecule which is raised to an excited state by inelastically colliding with a charged particle. The molecule then decays producing scintillation light. The pure chemicals are usually expensive and difficult to handle so they are dissolved in a plastic such as acrylic or an oil such as mineral oil to produce a useful scintillator. There are several types of organic chemicals that are commonly used for scintillators. Almost all of them are aromatics (benzene rings). Some examples are anthracine and p-terphenyl.

Not every collision between a molecule in the scintillator and an ionizing particle produces a quanta of light that is detected. The efficiency is the percentage of energy lost in the material which is converted to visible light. Typical efficiencies are around 22% for sodium iodide crystals[1], 6% for plastic scintillators such as NE110 and 4%[3] for mineral oil liquid scintillators. The wavelength where light output is maximum is typically around 400 nm.

Why are the efficiencies so low? First of all, not all collisions between a particle and the scintillator transfer enough energy to raise an electron into an excited state. Second, there are many competing mechanisms (especially in complex organic compounds) that allow a molecule to de-excite without emitting a photon. For example, the energy can be transferred into vibrational states of the molecules. Third, even if a photon is emitted, it may be absorbed by another molecule which may then de-excite by other mechanisms. The last mechanism gives rise to the attenuation length in scintillator. To increase the attenuation length most plastic scintillator has a wavelength shifter chemical added to it. These chemicals are chosen to have a strong absorption band at the wavelength of the scintillator and an emission band at a longer wavelength (lower energy). Much research has been done to find wavelength shifters which have no overlap between the absorption and emission bands so that the reabsorption probability is a minimum. Typical plastic scintillator attenuation lengths are a meter or two.

For single particle counting, scintillator efficiency is very important and the scintillator manufacturers as well as individual laboratories spend a lot of time optimizing this along with other features such as ease of manufacturing. However, loss monitors usually deal with such large numbers of particles that scintillator efficiency is not too important.

The other parameter in the amount of light produced per centimeter is the energy loss per centimeter (dE/dX). Minimum ionizing particles lose about 1.7 MeV/cm in plastic scintillators and 4.13 MeV/cm in sodium iodide. The above quantities are the most probable energy loss[7]. A useful rule of thumb for calculating the number of photons produced by a minimum ionizing particle is to use 100 eV of lost energy per produced photon for plastics or liquids and 25 eV per photon for sodium iodide. This is adequate for most loss monitor designs.

From the above data we see that sodium iodide crystals give off a lot of light - typically 160,000 photons per centimeter for a minimum ionizing particle. But they are fairly expensive and are difficult to handle. For example, the crystals are hygroscopic. They also have a very long decay time - typically around 230 ns. Thus they are rarely used in accelerator instrumentation. Their use in physics experiments is usually in electromagnetic calorimeters where their excellent light output leads to very precise energy measurements.

On the other hand, plastic or liquid scintillators are very easy to use. This ease of use more than makes up for the lower light output. Almost every high energy physics experiment has some plastic scintillators in its apparatus. Even though the light output per cm is only 1/8 that of sodium iodide, it is still 20,000 photons per cm which is more than enough to determine if a particle has passed through. The decay time constants are typically around 3 to 4 ns (70 times faster than sodium iodide)

Radiation damage is an important issue for some loss monitor applications. It is shown in the literature that most of the radiation damage occurs in the scintillating material itself rather than in the carrier. There are several good articles discussing radiation damage in detail so we won't discuss them here[8,9]. Most plastic and liquid scintillators can withstand up to 0.1 Mrad without significant damage. Above this amount the damage becomes very material dependent and one should consult the literature. Note that there is considerable evidence for fairly short term recovery of some of these materials.

Mineral oil based liquid scintillators are inexpensive and can easily be coupled to phototubes by simply immersing the tube in the oil. They are simple and easy to use and have found extensive use as loss monitors. One loss monitor developed at Fermilab[10] consists of a one gallon paint can (like you buy in a hardware store) filled with liquid scintillator. A hole is cut in the cover and a phototube is mounted in it. The cover holds the phototube, keeps the liquid scintillator in and makes a light tight box. The authors of reference 10 report that this system produced a 2 μA current for a 1 rad/hr dose rate.

What are the dynamic range and sensitivity of scintillators?   Unless they are physically damaged by radiation, scinilllators are quite linear and they have a very wide dynamic range. The linearity and dynamic range of the electronics is much more important .

Scintillators have been in use for many years and there is a well developed commercial market for them. The manufacturers can provide copies of many technical articles (the Nuclear Enterprises catalog[3] lists 49 technical papers) and advice on using their products.

Phototubes

The limit on linearity of scintillator loss monitors is the phototube. A schematic diagram of one is shown in figure 1. The basic principle of the device is to convert a photon to an electron by use of a photocathode. The physics of how this happens was first described by Albert Einstein in 1905 and the mechanism is called the photoelectric effect. If a photon is absorbed in a material has an energy greater than the work function of the material, then there is a good probability that an electron will be emitted from the surface of the material. The quantum efficiency of the best bialkalai photocathodes is around 25%. This means that for every 100 photons above a minimum energy, 25 electrons will be emitted.

After an electron has been emitted from the surface, it is accelerated to a structure called a dynode[11]. The acceleration is achieved by a potential of a hundred volts or so between the cathode and the first dynode. There are often focusing electrodes to improve the collection efficiency of the first dynode (F in figure 1). When the electron strikes the dynode, it has substantially more energy than the work function of the dynode material. Thus several electrons are emitted for each arriving one. This process is repeated for several more dynodes - perhaps as many as 14. Each dynode is held at a positive potential with respect to the previous one. If only three electrons are emitted for each arriving electron, then with 14 stages the final electron charge is 3^{14} or nearly 5 million electrons. Typical amplifications are closer to 1 million. Phototubes can thus be used for single electron counting.

Schematic Representation of Photomultiplier Tube (PMT)

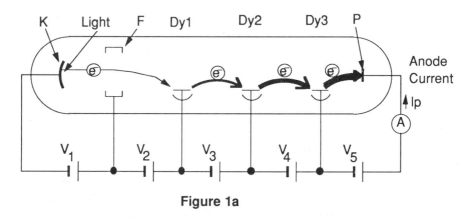

Figure 1a

Using Resistors only

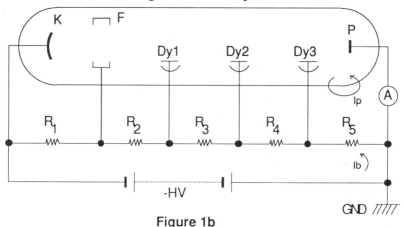

Figure 1b

Figure 1. Schematic diagram of a photomultiplier tube. Figure 1a shows a conceptual tube base using batteries while figure 1b shows a practical base using a resistor divider.

The simplest circuit for powering the dynodes of a phototube is a resistor chain connected to a high voltage power supply (figure 1). The resistor chain serves as a voltage divider to set each dynode at a specified voltage. This circuit works well for low numbers of photoelectrons. For high rates (the usual case for loss monitors) the current in the resistors for the last dynodes limits the gain. To see this, consider a current flowing from the last dynode to the anode. This movement of charge must be balanced by an equal amount of charge moving from the anode to the dynode through the last resistor. If the dynode to anode current equals that in the resistor chain, then there will be no accelerating voltage left and no signal. This effect can be reduced by connecting capacitors across the last 3 dynodes and the anode (in parallel with the resistors). This doesn't solve the problem if the input current is steady for periods greater than the RC time constant of the divider chain which may well be the case for loss monitors. This can be solved by using high voltage transistor amplifiers to drive each of the last few dynodes. The resistor divider then serves only to provide reference voltages. This type of base is more expensive and complex than one using resistors and capacitors so it is not used as often.

The next effect that limits the gain is space charge. As the cloud of electrons between the dynodes increases, the space charge reduces the accelerating voltage and thus reduces the gain. There is no simple method to compensate for this effect.

What is a typical signal from a loss monitor? As an example choose a 10 stage tube with a gain of 60,000 connected to a 5 cm liquid scintillator and a loss rate of 10^5 particles per second. From above we have about 14,000 photons per cm of liquid scintillator or 70,000 total for each particle. A 25% quantum efficiency coupled with a 25% collection efficiency then gives 4375 electrons per particle. Combining this with the gain and loss rate from above, we get 4.2 μa from the tube. The Hamamatsu[12] catalog states that the linear range for a typical resistor divider chain base goes up to 15 μA. For higher rates one can reduce the high voltage thereby lowering the gain or select a tube with fewer stages.. If losses are much smaller than this the voltage can be raised (but there is an upper limit) or one can go to a tube with more stages. The Fermilab paint can uses a 9 stage tube (RCA 931A).

Signal Processing

Modern electronics has for the most part solved the problems of signal processing. FET operational amplifiers can give high gain with low noise and 16 bit ADC's with 17 μsec conversion times are readily available. If more dynamic range is needed, there are logarithmic amplifiers with 5 decades of range and 100 KHz bandwidth and, of course, voltage to frequency converters.

One main use of loss monitors is to protect equipment from damage. Both rate and integrated dose may be important so many loss monitors are coupled directly to comparators or through integrators into comparators. The output of the comparator is then used to shut off the beam if it is triggered.

II. Gas Chamber Loss Monitors

Ionization of matter by charged particles can be exploited by using various types of gas based detectors. These devices have a long and varied history. The earliest device was the Geiger counter invented   by Rutherford and Geiger[13] in 1908. From this beginning many types of detectors have been developed but usually only the ion chamber and proportional counters are used for loss measurements.

Gas physics

Since the active material is a gas, one can develop a fairly good model of the ionization process. This is given by the Bethe-Block[14] formula shown in eq (1)

$$\frac{dE}{dX} = -\frac{2Dmc^2z^2}{\beta^2}\left\{\ln\frac{4m^2c^4\beta^4}{I^2(1-\beta^2)^2} - 2\beta^2\right\}, \quad D = \pi N\frac{Z\,e^4}{A\,m^2c^4} \tag{1}$$

where N is Avogadros number, m and e are the electron mass and charge, β is the incident particle's velocity as a fraction of the speed of light, z is its charge and c is the velocity of light. Z is the atomic number, A is the atomic mass and I is the ionization potential of the gas. From this equation one sees that the energy loss is independent of the mass of the incident particle but is a function of the particle's velocity. The ionization initially decreases with increasing velocity because of the $1/\beta^2$ term and then rises again as the log terms starts to contribute. Figure 2 shows data for ionization of argon by protons[15]. While eq (1) indicates a rapidly increasing energy loss as the incident particle approached the speed of light, figure 2 shows that this does not happen. As the ionization increases polarization effects come into play which are not included in (1).

Eq. (1) describes the average energy loss for a particle passing through a gas. It does not describe the distribution of energies that the ionized electrons receive. One might guess that this is a Poisson distribution since it is a random distribution associated with nuclear interactions. This is to some extent correct but it is not the whole story. To understand this we must differentiate between primary and secondary ionization. As you might guess, primary ionization is the ionization created by the primary particle passing through the gas. If the ionized electron is given enough energy so that it can

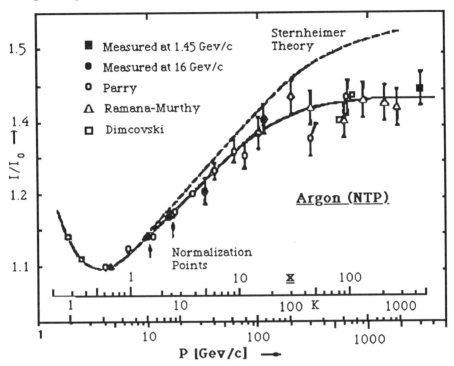

Fig. 2 Data from reference 15 showing ionization as a function of incident particle momentum for argon. The Sternheimer theory is an extension of the Bethe-Block formulation.

cause ionization (this electron is called a 'delta' ray) then one has secondary ionization. This secondary ionization is what causes the deviation from a Poisson distribution. The probability of transferring a large amount of energy to an electron is small but if it does happen, the pulse amplitude is large. Thus the ionization distribution is characterized by a Poisson distribution but with a long tail of large pulse heights. It is well described by a Landau distribution. Figure 3 shows data[16] collected with a streamer chamber. A streamer chamber sees all the ionized particles that are present at the time that the high voltage pulse arrives. At very short times only the primary ionization is present and ones sees in figure 3a that the streamer distribution fits a Poisson distribution. A short time later one sees the secondary ionization which fits a

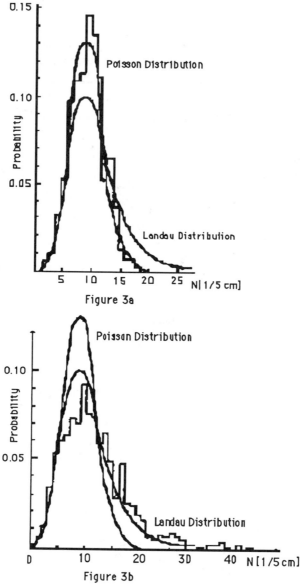

Figure 3. The distribution of the number of streamers per cm of track length in a chamber filled with helium at a pressure of 0.4 atm. The solid curves are Poisson and Landau distributions which have the same area as the data. Figure a has a time delay of 200 ns while figure b has a delay of 15 μ sec. From reference 16.

Landau distribution (figure 3b). Table I lists the number of ions per cm for primary and secondary ionization for several common gases.

Table I

Properties of several gases used in proportional and ion counters. Data is from reference 17. All data is at atmospheric pressure for minimum ionizing particles.

Gas	dE/dX (KeV/cm)	n primary (ion pairs)	n total (ion pairs)
H_2	.34	5.2	9.2
He	.32	5.9	7.8
N_2	1.96	(10)	56
O_2	2.26	22	73
Ne	1.41	12	39
Ar	2.44	29.4	94
Kr	4.60	(22)	192
Xe	6.76	44	307
CO_2	3.01	(34)	91
CH_4	1.48	16	53
C_4H_{10}	4.50	(46)	195

Most loss monitors use argon or a mixture of argon and CO_2. Both of these gases have a total ionization of about 100 pairs per cm of gas. Thus a good engineering rule of thumb is to use 100 ion pairs per cm of gas.

There are two broad categories of gas ionization detectors that are used for loss monitors; proportional counters which use gas gain and ion chambers which do not. Both types of detectors collect the free charges in the gas by putting a potential difference between two electrodes in the gas. The difference is that proportional counters use an electron avalanche in the gas to amplify the signal while ion chambers collect only the primary and secondary ionization. We will describe proportional counters first.

Gas Proportional Counters

The simplest type of proportional counter is the Geiger tube. It is made by stringing a small wire (about 20 μm in diameter) down the center of a tube (1 to 2 cm diameter), filling the tube with a gas such as argon and putting the cathode at a voltage around 1000-2000 volts. The field, E, at any point inside the tube at a radius, r, is given by:

$$E = V \frac{1}{r \ln(b/a)} \qquad (2)$$

where b and a are the radius of the tube and the wire respectively (figure 4). As one can see from this equation, the voltage becomes very large as one approaches the wire anode. When the voltage has increased to the point where the energy gained by a free electron in one mean free path of travel in the gas exceeds the ionization potential of the gas, then an additional electron may be created in the next collision. At close distances to the wire the number of electrons will double approximately every mean free path. It is quite easy to get a gain of 10^4 with this method. By adding small percentages of organic gases such as methane or ethane to the basic argon gas, gains of 10^6 can be achieved.

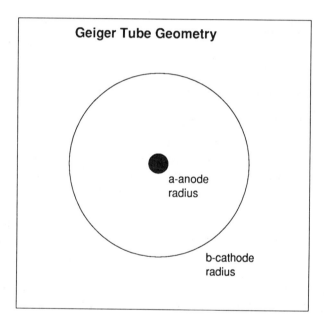

Figure 4. Geometry of a cylindrical Geiger tube. This drawing is not to scale! A typical ratio of cathode to anode radius is 1000:1.

The choice of gas for filling a detector is quite complex. Pure noble gases have a low breakdown voltage. This arises for two reasons. The first excitation level of argon is 11.6 eV so an excited argon atom can only decay by emitting an 11.6 eV photon (or by collision which only transfers the energy to another atom). This photon is considerably above the work function of any metal used for a cathode (the work function for copper is 7.7 eV for example). These photons can extract electrons over a wide region of the detector and start a wide area breakdown. The second reason is the result of the high ionization energy of argon (15.7 eV). An argon ion arriving at the cathode will have energy in excess of two times the work function of most cathodes. Thus they will either extract a second electron or else emit a photon. Both of these processes produce more free electrons.

On the other hand, polyatomic molecules such as CO_2 and methane have many additional rotational and vibrational states. Energy transfer from other excited atoms by either photon absorption or by collision is quite efficient. These complex, excited molecules usually either dissociate into subunits or undergo a multistep decay by emitting a number of soft photons which are below the work function of the cathode. Thus it is very common to add 10% to 20% CO_2 to a detector. If one is running at very high voltages for single particle counting, adding small amounts (0.5%) of gases such as freon-13B1 (CF_3Br) may be necessary. These gases attach free electrons forming a negatively charged ion. The negatively charged molecules must be accelerated to a much higher energy before they can cause ionization in collisions so the electron is effectively removed from the gas.

Adding organic gasses such as 10% ethane can increase the gain of a detector by more than an order of magnitude. However, these organic gases cause long term damage to the wire electrode thus limiting the life of the detector. The cause of this is thought to be the creation of free radicals from the breakup of the organic molecules by the ionizing radiation. These free radicals then react with each other and other parts of the detector to form coatings on the anode wires. Unless the increased gain is essential, it is best to avoid adding any organic gases. Also, if there is any possibility of outgassing organic compounds, it is best to continually flow fresh gas through the device. If the detectors must be sealed, it is best to use ion chambers which use a pure noble gas, run at low to moderate voltages and are constructed with only clean metal and glass. An example of this type of sealed detector is the Tevatron loss monitor described below.

Pulse Shape From Counters

It is worthwhile describing the operation of a proportional chamber in some detail and also developing the equation for the pulse shape. Consider one electron drifting to a wire in a Geiger tube. When it gets close enough to the wire, amplification starts. Assuming a doubling every mean free path, half of the signal is developed in the last mean free path before striking the wire - a distance of 0.4 μm for argon at atmospheric pressure and 20° C. Typical electron drift velocities are 10^7 cm/s but ion velocities are only 10^5 cm/s. Even when all the electrons have arrived at the wire, the positive ion cloud has not moved much at all. Thus the electrons are held by the positive ion cloud. I like to think of this using the old concept of flux lines for understanding the signal development. As the ions drift toward the cathode, the flux line to an electron on the anode is broken and then connected to one on the cathode. This allows an electron to flow off the anode and draws one onto the cathode. The shape of the pulse is almost entirely determined by the movement of the positive ions.

The time development of the pulse is important for detector builders so we will derive it in the next few paragraphs.[17] Note that this derivation is true for any cylindrically symmetric detector whether it be a gas chamber with gain, an ion chamber or a secondary emission monitor.

We will use the simplest system for this development, i.e., a cylindrically symmetric counter like a Geiger tube (figure 4). The electric field at any point between the anode and cathode is given by

$$E = \frac{CV_o}{2\pi\varepsilon_o}\frac{1}{r}, \quad C = \frac{2\pi\varepsilon_o}{\ln(b/a)} \tag{3}$$

where C is the capacitance per unit length, V_o is the anode to cathode voltage and a, b and r are anode radius, the cathode radius and the observation point respectively. This is a basic electric field law for cylindrically symmetric fields and it can be found in almost any text on electromagnetism.

From (3), one sees that if the inner radius, a, is small, the field becomes very large as the electron approaches the anode. When the energy gained in one mean free path exceeds the ionization potential of the gas one gets amplification. Since the mean free path is typically 0.4 microns and 13 doublings give one a gain of 8,000, all the amplification occurs within a few microns of the anode.

Eq. (3) gives the electric field in the Geiger tube and we want the potential. This is given by the integral of (3) from the inner radius, a, to the observation point. If the observation point is the outer radius, b, then it must give the applied potential V_o.

$$V = \int E\ dr = \int_a^r \frac{CV_o}{2\pi\varepsilon_o} \frac{1}{r'}\ dr' = \frac{CV_o}{2\pi\varepsilon_o} \ln\frac{r}{a} \tag{4}$$

Now we need to find the change in the voltage on the detector when a small charge is moved inside the tube. That is, when we move a charge inside the detector, the voltage across the detector will no longer be V_o but rather V_o plus a small induced voltage. This can best be derived by looking at the change in energy of the system. The energy of a clump of charge Q at potential V is simply Q V. Think of a capacitor. The work required to move a charge from the ground side to the high side of a capacitor is just the amount of charge times the voltage on the capacitor.

Since the Geiger tube has capacitance, there exists a charge on it even when there are no free electrons inside. Let this charge be called Q. Let Q' be the free charge in the tube and dv be the output pulse. We want to calculate dv (the induced voltage).

The energy associated with charge Q' at position r in the tube is Q'V(r) where V(r) is the potential at the charge position. The change in energy of this charge in moving it a distance dr is then

$$dE' = Q'\frac{dV(r)}{dr}\ dr \tag{5}$$

The energy change in the entire tube is given by $dE = Q\ dv$. These two energy changes must be equal so we get

$$Q'\frac{dV(r)}{dr}\ dr = Q\ dv \tag{6}$$

or

$$dv = \frac{Q'}{ClV_o}\frac{dV(r)}{dr}\ dr \tag{7}$$

since $Q = ClV_o$. Here l is the length of the detector. Thus we have the induced signal in terms of the rate of field change where the charge is. To get the total pulse, v, we integrate (7) from the point where the charge was created, r', to the point where it is collected (the anode a). This gives

$$v = \frac{Q'}{CIV_0}[V(r')-V(a)] \tag{8}$$

Even though we have used the Geiger tube as an example, there is nothing in the derivation so far that relied on that geometry - the above result is completely general. It also works for secondary emission detectors.

For cylindrical geometry, substituting V from (4) gives

$$v = \frac{Q'}{CIV_0}\frac{CV_0}{2\pi\varepsilon_0}\ln\frac{r'}{a} = \frac{Q'}{2\pi\varepsilon_0 l}\ln\frac{r'}{a} \tag{9}$$

As an exercise we will compute the ratio of the signal of the electrons to the ions in a cylindrical Geiger tube. More than 90% of the charge is generated in four mean free paths of the anode which is about 1.6 microns. The signal from the electrons is

$$v^- = -\frac{Q'}{2\pi\varepsilon_0 l}\ln\frac{a+1.6}{a} \tag{10}$$

and the signal from the ions is

$$v^+ = \frac{Q'}{2\pi\varepsilon_0 l}\ln\frac{a+1.6}{b} \tag{11}$$

Typical values are a =10 and b = 6000 so the ratio v^-/v^+ = 0.023 so only about 2% of the signal comes from the electrons.

We want the time development of the pulse so we need to find r' as a function of t and then substitute into (9). For all ions and electrons in some gases the drift velocity is linearly dependent on the applied electric field, E, and inversely proportional to the gas pressure, P.

$$\frac{dr}{dt} = \mu\frac{E}{P} = \mu\frac{CV_0}{2\pi\varepsilon_0 P}\frac{1}{r} \tag{12}$$

where μ is the drift constant. We can solve (12) for r which gives

$$r \, dr = \mu \frac{CV_o}{2\pi\varepsilon_o P} \, dt \qquad (13)$$

Assume ions drift from anode to cathode. Integrating (13) from the inner radius a (t=0) to position r' (t=t') and solving for r'(t') gives

$$r'(t') = \sqrt{a^2 + \frac{\mu CV_o}{\pi\varepsilon_o P} t} \qquad (14)$$

Finally, substituting this into (9) gives

$$v(t) = -\frac{Q'}{2\pi\varepsilon_o l} \ln\left(1 + \frac{t}{t_o}\right) \text{ where } t_o = \frac{\pi\varepsilon_o P a^2}{\mu CV_o} \qquad (15)$$

where l is the length of the chamber and t_o is a characteristic time. v(t) is the change in voltage across the tube caused by the movement of the positive ions.

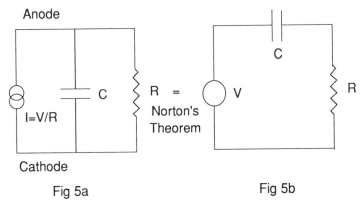

Fig 5a   Fig 5b

Figure 5. Equivalent circuit for a gas chamber detector

Figure 5 shows an equivalent circuit for a counter. Since the charge is driven by the applied high voltage, it is nearly a true current source. The capacitor, C, is the capacitance of the tube and R is the anode-cathode impedance as seen by the output signal. If R is infinite, then the charge will not recombine and there will be a permanent voltage change across the tube given by v(t) (see the example below).

The circuit of fig 5b can be solved as follows.

$$\frac{1}{C} \int i(t)\, dt + R\, i(t) = v(t) \quad \text{or} \quad R\,\frac{di(t)}{dt} = \frac{dv(t)}{dt} - \frac{i(t)}{C}. \tag{16}$$

If R is zero, then the current is

$$i(t) = C\frac{dv}{dt} = \frac{Q}{4\pi\varepsilon_o t_o l}\, \frac{1}{\left(1 + \dfrac{t}{t_o}\right)} \tag{17}$$

If R is finite, then we need to solve (16). Unfortunately (16) cannot be solved in terms of elementary functions (it involves an exponential integral). However, in most cases R is small (50 ohms or so) so the voltage is still differentiated and we have

$$v_o(t) \approx R\,\frac{Q}{4\pi\varepsilon_o t_o l}\, \frac{1}{\left(1 + \dfrac{t}{t_o}\right)} \tag{18}$$

for the pulse shape from a Geiger tube.

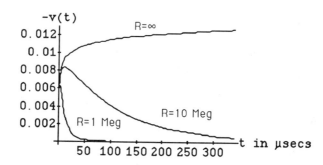

Figure 6. Output voltage as a function of terminating resistor value. The vertical axis units are arbitrary.

Let us work an example using the parameters mentioned above (b=6 mm, a=.01 mm and l= 1 m). Using the above formulas we get C=8.7 pf/m. From Sauli (1977) μ for argon ions is 1.7 cm^2 sec^{-1} V^{-1} atm^{-1}. The total drift time using (14) with r=b is 340 μsec and t_o=1 nsec. Figure 6 shows the voltage pulse for a Geiger tube for 3 different values of R. One sees that as the resistance gets smaller the signal comes closer to being differentiated.

Figure 7 shows the current pulse, i(t), for the cases R=0 and R=50 Ohms. The two curves have very similar shapes. A detailed comparison shows differences of order 20% at the peak which means that 50 ohms is not entirely negligible compared to the the impedance of the series C (fig 5b).

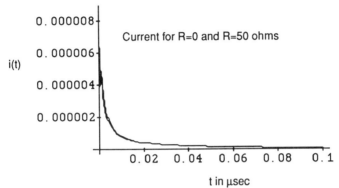

Figure 7. Output current for R=0 and R=50 ohms. The vertical axis units are arbitrary.

Thus the time evolution of the pulse follows a $1/(1+t/t_o)$ distribution.

If the anode is composed of a parallel wire grid with spacing about 1/3 of the anode to cathode spacing and the cathode is composed of two parallel plates or grids, then the field will be uniform throughout most of the gap but will follow (2) as a wire is approached. Fig 8 shows a plot of the equipotential lines[18]. The uniform field will drift the electrons to the closest wire for amplification. The struck wire then provides position information. This type of detector has been highly developed during the last 20 years. Several good references are listed in the bibliography.

Loss monitors using gas gain are not as common as ion chambers. They are typically used where the beam is very clean or of low intensity. Since they have gain, they can use a smaller volume of gas for the same signal and they can have a faster rise time. Gas gain can also reduce the required electronic gain which may make the device's amplifiers much easier to construct.

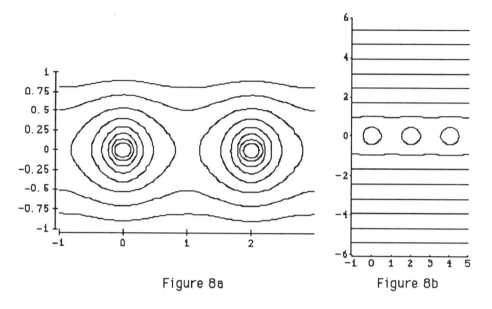

Figure 8a   Figure 8b

Figure 8. Equipotential lines from a chamber with anode-cathode spacing of 6 mm, an anode wire spacing of 2 mm and an anode diameter of 10 microns. Figure a shows a close up of the field around the wire while figure b shows the overall field.

Ion Chambers

Detectors that do not use gas amplification are called ion chambers. These devices collect only the primary and secondary ionization produced by the particle traversing the chamber (100 electrons per cm of argon) so the signal is extremely small. Since they have no gas amplification, they can run at low voltages (a few hundred or so volts per centimeter) and they do not require a small diameter anode. The anode and cathode can be simple parallel plates for example. Thus they are simple and rugged. They also have very large dynamic range. Since each particle produces only 100 ion pairs per cm, it takes an enormous flux to put the device into the non linear region by fully ionizing the gas.

The signal development from an ion chamber can be quite different from that of a gain chamber. Since there is no gain, the signal from the ions is almost always important. In the limiting case where the ionization is formed next to the cathode, all the signal is from the ions. This also means that electronegative gases such as oxygen have little effect except on the signal rise time (see below). A negatively charged ion has about the same drift velocity as a positively charged one. Thus some people have successfully made ion chambers that work on normal air.

A loss monitor using atmospheric pressure argon gas was built at Fermilab for the Tevatron[19] It consists of a cylindrical glass tube and nickel electrodes (see figure 9). Since the gas is sealed inside, the designers were very concerned about radiation induced aging. That is why pure argon, glass and nickel were chosen as materials.

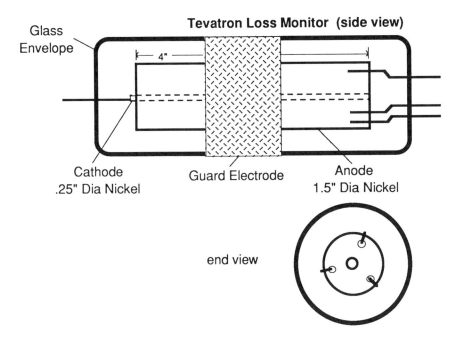

Figure 9. Schematic of Tevatron loss monitor. The monitor is filled with argon gas at 725 mm of Hg. The guard electrode reduces the leakage current to about 10^{-14} amps.

Figure 10 shows the peak output voltage of the log amplifier as a function of radiation dose[20]. A log amp was chosen to give the large dynamic range that was required. The devise is calculated to be linear to 100 rads instantaneous dose rate. This corresponds to 7 μcoulombs of collected charge. The lower limit is set by electronic noise and leakage currents in the device but a dynamic range of at least 10^6 is achieved.

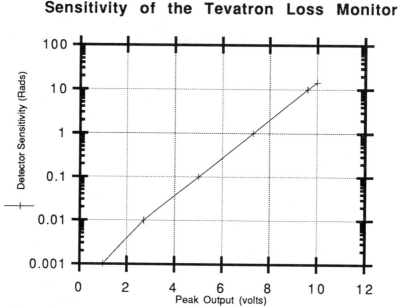

Fig. 10. Measured response of the Tevatron loss monitor as a function of the radiation input to the detector.

Since the electron drift velocity is about 100 times that of the ions and for argon and CO_2 it is proportional to the applied voltage, one can get a fast electron signal by running the tube at high voltages. A field of 1200 V/cm gives a drift velocity of 0.8 cm/µs. The drift distance in the Tevatron loss monitor is around 1.5 cm so the electron signal is obtained in 1.9 µs or less which is 5 to 10 times faster than the ion signal. This fast signal is used for beam aborts to prevent quenching the superconducting magnets.

Loss monitors have also been made using hollow core coaxial transmission line. This design can literally be strung for hundreds of feet along linacs or beam extraction areas. All that is needed is an RF connector and gas fittings. They were originally developed at SLAC to measure losses along the linac[21]. They can be hundreds of feet long and since the detector is very high quality coax cable, one can measure position as well as intensity by measuring the arrival time of the pulse at the end of the cable. They report a position resolution of 5 feet with a sensitivity of 20 rad/hr. Note that for this to work, the beam pulse must be very short since they measure the ion chamber pulse arrival time with respect to the beam pulse. Similar chambers used for a short proton linac may not do nearly as well. A system built at Brookhaven Laboratory[22] (using 7/8 inch HJ5-50 Heliax) solves this problem by chopping up the coax into short lengths (about 25 feet) and using the entire detector as a position locator. This costs a little more in electronics (BNL used 32 sections) but eliminates the dependence on beam pulse width.

These devices contain a lot of plastic which may outgas with time and radiation dose. This may not be much of a problem from the point of electron attachment but these complex gases may polymerize under radiation and coat the electrodes thus reducing the efficiency. This type of radiation damage is well known for proportional counters that use a mixture of organic gases. Thus, it is advisable that gas be flowed through these system.

Most loss monitor systems use some sort of log amp to get the required dynamic range. These devices should be located as close to the detector as possible in order to minimize the possibility of noise pickup. It is also best to use a single point ground with the HV power supply located close to the detector electronics.

Multi-Anode Detectors

Most loss monitors have only one anode. But it is quite easy to have several anodes in the same gas volume. The electronics cost for preamplifiers is often much less than the associated mechanics and high voltage so this may be cheaper solution to increased resolution than multiple detectors. Alternatively, a multi-anode monitor may give additional position information at little additional cost. The resolution (sigma) of a multiwire proportional chamber is the wire spacing/$\sqrt{12}$. Wire spacings of 2 mm are easy to construct so quite precise measurements can be made. Multi anode ion chambers can also be made. Their resolution depends on the distribution of charge among the different anodes which is very dependent on the detector geometry.

III. Semiconductor Loss Monitors

There has been a long history of using solid state devices as detectors for experimental physics. Work has also been done to develop silicon loss monitors[23]. This system is similar to a solid state TV camera. The intrinsic layer between the p and n type materials is almost free of charge carriers when the diode is reverse biased. Ionizing radiation creates electron- hole pairs in this region which cause a current to flow. The current is directly proportional to the amount of radiation present. The diodes are also very fast so one can get a good time distribution of the losses. The efficiency is reported to degrade by 6% per Megarad for doses up to 5 Megarads. The only draw back is the small size. Typical devises have active regions of only 1 mm^3.

IV. HEP Detectors As Loss Monitors

It is now becoming common to locate experiments inside the accelerator complex itself. The goal of running the accelerator may be to provide beam for these experiments. In this case, minimizing the loss at these detectors may become a prime goal. These detectors will provide a far better loss monitor than one could reasonably construct so one should try to use them. Unfortunately, these devices are not set up as loss monitors and the data they provide may be difficult to use to tune the beam. A system for the final focus at SLAC[24] used the MARK II detector. Initially the data arrived with a time delay of some 30 seconds which made beam tuning difficult. Additional hardware was then added and some software changed to improve this time.

V. Bibliography

Books On Detectors

M. A. Cook et al, editors, *Liquid Scintillation Counting, Vol 2,* Proceedings of a Symposium on Liquid Scintillation Counting, (1971), Heyden & son, London

T. Ferbel, editor, *Experimental Techniques in High Energy Physics,* Addison-Wessley, Menlo Park, CA.

R. Fernow, *Introduction to Experimental Particle Physics,* Cambridge Univ. Press, New York (1986).

P. Rice-Evans, *Spark, Streamer, Proportional, and Drift Chambers,* Richelieu Press, London.

Serge A Korf and H. Kallmann, *Electron and Nuclear Counters; Theory and Use*, Greenwood Press, New York, 1955.

Articles On Radiation Damage To Detectors

P. H. Levy, *Overview of Nuclear Radiation Damage Processes: Phenomenological Features of Radiation Damage in Crystals and Glasses*, published in Radiation Effects in Optical Materials, SPIE vol 541,(1985).

G. Marini et al, *Radiation Damage to Organic Scintillation Materials*, CERN technical report 85-08, June, 1985

G. Oldham and A. R. Ware, *Gamma-Radiation Damage of Organic Scintillation Materials*, J. Inst. Nuc. Eng., **12**, 95-97 (1971).

M. Ramakrishna Murthy and S. Radhakrishna, *Radiation Damage in Some Synthetic High Polymers*, Indian J. Pure & Applied Phy., **21,** 582-586 (1983)

I. M. Rozman and K. G. Zimmer, *The Damage to Plastic Scintillators by Ionizing Radiation*, Inter. J. Applied Rad. & Isotopes, **3** 36-42(1958)

H. Schonbacher, *Review of Radiation Damage Studies on Scintillating Materials Used in High Energy Physics Experiments*, CERN technical report TIS-RP/201, November 1987

H. Schonbacher and W. Witzeling, *Degradation of Acrylic Scintillator and Wavelength-Shifter Material by Nuclear Radiation*, Nuc. Inst. & Meth. **165**, 517-521 (1979)

Y. Sirois and R. Wigmans, *Effects of Long-Term Low-Level Exposure to Radiation as Observed in Acrylic Scintillator*, Nuc. Inst. & Meth., **A240**, 262-274 (1985)

Articles On Beam Loss Monitors

V. Agoritsas et al, *A Microprocessor-Based System for Continuous Monitoring of Radiation Levels Around the CERN PS and PSB Accelerators*, Nuc. Inst. & Meth., **A247**, 44-49 (1986).

M. Awschalom et al, *An Inexpensive Loss Monitor for Use at NAL*, Fermilab technical report TM-274, October 1970.

R. G. Jacobsen and T. Mattison, *Beam Monitors in the SLC Final Focus*, Proc. 1989 IEEE Part. Accel. Conf, Washington D.C.

G. S. Levine and J. C. Balsamo, Jr., *The AGS Beam Loss Monitoring System*, BNL Report 19836.

R. D. Raffnsoe, *Silicon PIN Diodes as Radiation Detectors*, CERN Report, August, 1983.

T. C. Ries et al, *An External Monitor of Beam Loss in an H⁻ Cyclotron*, Proc. 1987 IEEE Part. Accel. Conf, Washington D.C.

R. E. Shafer et al, *The Tevatron Beam Position and Beam Loss Monitoring System*, Int. Conf. of High Energy Accelerators, Fermilab 609 (1983).

R. L. Witkover, *Microprocessor Based Beam Loss Monitor for the AGS*, IEEE Trans. Nucl. Sci. **26** 3313 (1979)

References

[1] Korf,Serge A. and H. Kallmann,*Electron and Nuclear Counters; Theory and Use*, Greenwood Press, New York, 1955.

[2] Kleinknecht, K. *Experimental Techniques in High Energy Physics*, ed by T. Ferbel, Addison-Wessley, Menlo Park, CA. p 1.

[3] Nuclear Enterprises Scintillator Catalog, 931 Terminal Way, San Carlos, CA 94070

[4] A good introduction to the different mechanisms of producing light can be found in K. Nassau, *The Causes of Color*, Scientific American, **243** 4 (October, 1980) p. 124-154.

[5] R. L. Heath et al., Nuclear Inst. Methods, **162**, 431 (1979).

[6] M. C. Ross et al., *High Resolution Beam Profile Monitors in the SLC*, Proc. 1985 IEEE Part. Acel. Conf., Vancouver, B.C., Canada.

[7] M. Aguilar-Benitez et al, *Review of Particle Properties*, Phys. Lett. **170 B** (April, 1986)

[8] G. Marini et al, *Radiation Damage to Organic Scintillation Materials*, CERN technical report 85-08, June, 1985.

[9] H. Schonbacher, *Review of Radiation Damage Studies on Scintillating Materials Used in High Energy Physics Experiments*, CERN technical report TIS-RP/201, November 1987

[10] M. Awschalom et al, *An Inexpensive Loss Monitor for Use at NAL*, Fermilab technical report TM-274, October 1970.

[11] A good description of photomultiplier tube circuits and their limitations can be found in *Accessaries for Photomultiplier Tubes*, Hammamatsu Photonics co.,420 South Avenue, Middlesex, NJ 08846

[12] Hammamatsu Photonics K.K., Electron Tube Division. U. S. sales office: 420 South Avenue, Middlesex, N.J. 08846. They have several catalogs on phototubes and their accessories.

[13]E. Rutherford and H. Geiger, *Proc. Roy. Soc.* ,**A81**, 141 (1908).

[14]P. Rice-Evans, *Spark, Streamer, Proportional, and Drift Chambers,* Richelieu Press, London.

[15]D. Jeanne et al, Nuclear Inst. Methods, **111**, 287 (1973).

[16]V. A. Davidenko et al, Nuclear Inst. Methods, **67**, 325 (1969)

[17]F. Sauli, *Experimental Techniques in High Energy Physics,* ed by T. Ferbel, Addison-Wessley, Menlo Park, CA, p. 131.

[18]G. A. Erskine, *Nuclear Inst. Methods,* **105**, 565 (1972)

[19]R. R. Shafer et al, *The Tevatron Beam Position and   Beam Loss Monitoring System,* Int. Conf. of High Energy Accelerators, Fermilab 609 (1983).

[20]R. Shafer, Fermilab internal note.

[21]R. G. Jacobsen and T. Mattison, *Beam Monitors in the SLC Final Focus,* Proc. 1989 IEEE Part. Acel. Conf., Chicago, IL.

[22]R. L. Witkover, *Microprocessor Based Beam Loss Monitor for the AGS,* IEEE Trans. Nucl. Sci. **26** 3313 (1979)

[23]R. D. Raffnsoe, *Silicon PIN Diodes as Radiation Detectors,* CERN Report, August, 1983.

[24]R. G. Jacobsen and T. Mattison, *Beam Monitors in the SLC Final Focus,* Proc. 1989 IEEE Part. Acel. Conf., Chicago, IL.

BROOKHAVEN INSTRUMENTATION WORKSHOP ROUNDTABLE

Moderator: *Olin B. van Dyck*
Los Alamos National Laboratory

INTRODUCTION:

This is the edited transcript from the Roundtable discussion held the first night of the Workshop. Panelists representing several accelerator laboratories, associated businesses, and the USDOE covered many topics including the typical costs of accelerator instrumentation, standardization within and among laboratories, development by industry, and the USDOE Small Business Innovative Research support program.

In the background was the question whether more standardization would save resources at the laboratories, and presuming so, what factors have inhibited commercial development as extensive as in particle detector electronics for experimental physics. In the discussion, we heard that many factors may contribute, including both technical factors such as unique requirements, and administrative notions such as the need to maintain a strong in-house tech base. Countering these forces may be the realization of the true costs of stand-alone development, willingness of the DOE to support commercial ventures, and the possibility of de facto standards set by the SSC.

The transcript was extensively edited by the moderator, with many paraphrases and abridgements, to meet the requirements of grammar and coherence. The voices had to be identified by the moderator from the recording tape, and I apologize for the inevitable misattributions and misinterpretations.

ROUNDTABLE

VAN DYCK: Let's introduce ourselves:

SHAFER: I'm Bob Shafer from Los Alamos National Laboratory. I'm a physicist in the Accelerator Technology Division.

ROSSMANITH: My name is Robert Rossmanith. I'm with CEBAF, the Continuous Electron Beam accelerator Facility, and I'm in charge of beam instrumentation and control.

HUBBARD: I'm Rex Hubbard from the Jorway Corporation.

PRIESTER: I'm Don Priester from the Department of Energy, with the office of SSC.

ODUM: My name is Robert Odum. I work at Charles Evans & Associates and I'm involved with some SBIR programs.

VAN DYCK: I'm Olin van Dyck from LAMPF; I'm a facility development physicist.

LAUTERBACH: I'm Mike Lauterbach from Lecroy Corporation. I'm Vice President there. Perhaps it is worth mentioning that I used to work at Brookhaven and Fermilab before that.

REZZONICO: I'm Luigi Rezzonico from Paul Scherrer Institute, previously SIN, in Switzerland, and I'm responsible for all beam diagnostics.

WEBER: I'm Bob Weber from Fermilab. I work in the instrumentation group and I have been primarily involved in beam position and beam intensity measurement instrumentation.

HINKSON: My name is Jim Hinkson from LBL in Berkeley. I'm lead engineer on the instrumentation and low level electronics on the ALS project.

ROSS: I'm Mark Ross from SLAC. I'm an accelerator physicist involved in many general aspects of SLC.

WITKOVER: I'm Richard Witkover from Brookhaven and I'm responsible for the instrumentation for our new booster project.

VAN DYCK: Thank you gentlemen and welcome to the first annual instrumentation roundtable. I would like to address a comment and a question first to Dick Witkover. At Los Alamos we have a half mile of Linac with wire scanner system A. The Linac then feeds the switchyard with wire scanner system B which has a different style of actuator, different electronics, different computer system, and different software and then the beam goes into another beam line where it has yet a third system of wire scanners. A conclusion that I would tentatively draw from that is that even within one accelerator people so much want to have the instrumentation under their own control if they are responsible for that area of the facility that they prefer to do it themselves. So a possible explanation for why their is so little interlaboratory cooperation and so little commercialization in instrumentation is that we are actually not fighting a technical or economic problem, we are simply fighting a human or management problem. Dick, you have worked on organizing this workshop for two years so you must believe there's an answer to that.

WITKOVER: I wish I could say there was an answer. That's probably the hardest problem we have to solve. Certainly at Brookhaven we have the same situation, to use your example, with several different types of wire scanners. I intend that in the future, certainly on the booster, that will not be the case. We have three transport lines and since there was a coordinated effort to put them in, they are all the same; they are also the same as the ones that we had previously installed. I can speak for that because I had some control over it; I cannot speak for what will happen in the future unless I'm involved with it. I think that it takes encouragement from a higher authority, if you will, to force this to be. We cannot keep on making our own little empires and doing this type of thing. We shouldn't always be reinventing the wheel but enhancing the wheel: to go from one to the other with evolutionary steps rather than totally independant revolutionary steps. As I said before to a number of you if we had more of a common ground industry could support us. To give you an example, these same wire scanners are ones that I purchased commercially, in fact from a vendor representing a company not in this country. We do not have that type of an organization here to produce these things and have a viable market because we all do our own thing. That is one of the problems that you very well brought up, Olin, and it is typical not only to Los Alamos but I daresay everywhere else everyone has their own organization. Maybe even in the next building or the next office there is another one that is doing the same thing but differently and I certainly have not the solution for it.

VAN DYCK: Bob Weber, as a representative of one of the world's leading accelerator laboratories, perhaps we can ask you how big a part of the Laboratory's capital plant and how big a part of its operating expenses is your I & C? Do you think a standardized solutions would have helped you or are you big enough to justify building and supporting your own self anyway?

WEBER: Well, in regard to the first question as to the capital investment and operating costs of the instrumentation effort, let me separate that from controls. In each of the accelerators, the Tevatron and the booster, main ring, and the antiproton source rings, the beam position systems are probably in the range of a half million to two million dollars, the largest expense being in the Tevatron. The main ring is very comparable and the booster and the antiproton source are somewhat less. Historically what happened at Fermilab was that for the booster and the main ring there was this new idea--someone came along and said we needed the beam position system in the main ring and the system was built. That was in the early seventies. In the booster system, which was built earlier, the beam position detectors were installed when the machine was built and some electronics which worked quite miserably was put in. A great deal of respect was given to the answers that came out and we suffered from that. But the systems we have now are really all offsprings of the system that was developed for the Tevatron. I guess because we were in the position of building essentially the largest and most complicated accelerator that had been built to date, there was sufficient money put in it to make sure that system worked. There was sufficient time available because the superconducting magnets were such a new technology that they were the longest lead item. So I think we did at that point in the Tevatron have the luxury of time to be able to think and plan what was necessary. Bob Shafer did a lot of this and we came up with a system that was probably the largest and most complicated beam position system that has been put in an accelerator at that time. In that sense we were breaking new ground and I'm not sure what sort of standardization would have been useful to us. Maybe one is being selfish in saying "look, we are the biggest so everybody else should do what we do." There certainly are cases where that might work and many cases where it simply will not work. But within Fermilab, maybe because there were a few people in the same positions of power for a long enough period of time, we learned the value of the Tevatron beam position system, including all of the various built-in test capabilities and so on that Bob Shafer mentioned this afternoon. After recognizing that, we said, "well, let us adapt that system to work in the main ring." We had the fortunate circumstance that the beam properties in terms of the bunching, frequency range, intensity range, and all of this were vary comparable because the main ring feeds the Tevatron. So the system was essentially duplicated in the main ring. Then we simply worked backwards, but in the booster instead of working primarily at a single frequency of 50 MHz we had to cover the 30 to 50 MHz band. We adapted the beam position electronics with the addition of a frequency conversion stage at the input and essentially that was the only difference from the Tevatron beam position electronics. Later, the Linac people got jealous, I guess, and wanted a beam position system so we adopted the frequency conversion stage to 200 MHz input. We used that economy to get the job done with a minimal amount of effort. That's the analog end.

The problems in this standardization scheme or the stretching thereof came when we tried to adapt the controls system end and the local intelligence section of the beam position system of the Tevatron. On the main ring it was not so difficult. We did not need some of the features that had been put in for the Tevatron, for instance the abort capabilities which were there to protect the Tevatron superconducting magnets. So that-back pedaling to transform the system for the main ring was rather painless. But when we put the same system into the booster we ran into real problems after the analog front end. Some of the complications are still a "burr under the saddle" today. They include things that are related to the revolution time in the booster ring being on the order of a microsecond whereas in the Tevatron it is about 20 microseconds. When we finally got to the Linac, even with the idea that we should keep things standardized, it became obvious that the local intelligent microprocessor system simply was not suitable for single pass operation. There we went to a much simpler sampling system. Maybe we carried standardization to the limit and maybe to the point of being painful just to maintain the same hardware. You can stretch it beyond anything that is reasonable.

AUDIENCE (unidentified): From what I have seen in my years at Fermilab and other labs, everyone has a very different control structure and you get into the control structure when you have to build things to adapt to the way we have already done it here. When new architectures like VME come along we hear, "sorry, we are not going to support that because we do not do that here." Trying to standardize within the laboratory is tough, amongst multiple laboratories is tougher still because of this attitude. You are kind of stuck with the architecture that your laboratory adopted from day one.

WEBER: That may be a result of the fact that the capital investment in the control system is so much larger than the investment in the instrumentation equipment.

SHAFER: Perhaps I can make a comment on this since I have been at Fermilab probably more than most other people here. When Wilson set up Fermilab he actually set up separate groups for each machine and experimental area, that is, linac, booster, main ring, and three experimental areas. Each group was autonomous and each had their own concept as to how to build a control system. Some of you people probably know the guy who did the control system for the booster before he left and founded Kinetic Systems. His Nimbin, for example, had +/- 15 volts and +5 and a different Nimbin connector in it so you really could not plug in standard modules. Every control system for every accelerator was different and it was a real hodgepodge. Within Fermilab even though these groups started up at the same time they all had their own individual concepts as to how to build a control system. It really suprises me that Wilson did not force a Fermilab-wide control system group but let each of the experimental areas and accelerators be autonomous in their controls.

VAN DYCK: I think it has often been difficult for managers to understand how expensive this proposition was becoming. They tended to accept advice from managers individually, who said, "you give me a job to do you tell me the specifications on that job and I'll do it but do not tell me how to do that job." A manager under those circumstances is very reluctant to force a particular technical solution on to a technical person.

AUDIENCE (unidentified): At LAMPF, Fermi, and Brookhaven almost all of the experimental areas are supplied by equipment that is manufactured by outside firms. Lecroy, Phillips, and others supply almost all of the experimental electronics. You might ask why in those areas does it work?

SHAFER: Would you like me to give you an answer right now to that question, how we handled it at Fermi? In the early days at Fermi when we started supplying equipment to experimenters, each experimenter wanted his own kind of equipment. Some demanded Chronetics, for example, which came in a non-NIM module kind of thing and was hot enough to cook an egg on and was very different from some of the other electronics. It was a very astute decision not to give these users the flexibility of buying their own equipment with money that we provided. Instead, the money supported a group called **PREP** which I ran in the early days and which would negotiate with the users to provide them with the functions they wanted in the equipment. We very carefully wrote specifications for the equipment which satisfied all their needs and we convinced them that this is what they needed rather than what they desired. We got a commonality of all the systems they needed and we went out for competitive bids from **PREP** not on behalf of the individual users but from **PREP** which was a laboratory organization. We went out to small companies, for example, in West Nyac which is now Lecroy, and to EG & G and others who at that time were competitive in this business; and also Chronetics which was on its last legs--I do not know what has happened to them since. We went out for very large orders. We were buying fast electronics at 2 million dollars a year for awhile and providing it to the experimenters but we forced the commonality of this equipment on the users. Basically we told them you can have what you want provided you use what we give you. It was basically getting the vendors to produce equipment that met our specifications which were really quite rigid so it was in essence forced on the users to use this common equipment. We convinced them that relinquishing their individual choices to have individual equipment allowed us to have a bigger support pool of common equipment which they could all draw on if they needed it. And it was this commonality and availability of the pool that led us to purchase these big quantities from single vendors.

AUDIENCE (unidentified): The comment I would like to make though is that there is a big difference between a physics experiment and an accelerator. Anyone who has seen a physics experiment will know it is just a big spaghetti mess of wires and lasts from anywhere from six months to two years but if you build an accelerator you make it to last for many experiments. An experiment is a real kluge; it is poorly documented and there are wires all over the place. So when you make an accelerator that has to last any length of time you start saying, OK, instead of having twenty thousand wires attaching to these CAMAC cards let's make something custom that may even be goof proof so an operator can replace something in the middle of the night.

SHAFER: Seems to me the early days in the main ring, our RF building was kind of a kluge too!

VAN DYCK: Let me continue along the topic of how much an individual accelerator laboratory is able to standardize. I would like to address my question to Jim Hinkson. Jim represents a laboratory that has been around a long time and has a very distinguished history and supports a variety of accelerators and beams.

HINKSON: We invented physics, guys! For the Advanced Light Source project the controls systems group is going to spend about four million dollars on the construction effort. I believe we are in the second year now of a five year job for instrumentation and low level electronics and that includes a whole bag of things. We are talking about three and a half million dollars and in the area of beam position monitoring it is about a million and a half. There is a particular emphasis on beam position monitoring for the storage ring; it has some unusual requirements. Now I would like to address this standardized solution thing a little bit. Historically at the Lab whenever something wasn't available, we built it, but over the years we have seen standardized equipment appearing from industry. For example, only fifteen years ago I would build my own ion gauge power supply but now I can go to maybe four manufacturers--Varian, Perkin-Elmer, Altec or whomever. For transductors we always built them; now we buy them. For fast scopes we used to build our own until Tektronix got on the ball. We had a division of people devoted to just developing counting instruments for nuclear physics experiments; now we can buy these things from Lecroy, ORTEC, and the like. High voltage power supplies-- there were no manufacturers of reliable high voltage power supplies; now we have a number of outfits to choose from. Fibre optic systems--fifteen years ago we built our own fiber optic systems to span 750 kV; now we can buy them. Valves--we can buy ultra high vacuum valves now that we used to have to build. Fast valves--we used to build them, now we buy them. I see standardization coming on as there is an industry-wide need for these things and suddenly they're out there. So we will seek a standard solution where we find qualified components, stuff that works and we don't have to build it. We still have a committment to in-house engineering and expertise on certain topics but standardization is at work and we do take advantage of it.

VAN DYCK: So you see standardization creeping in not by concerted effort but as a response.

HINKSON: I think these manufacturers have to make a buck and they look at the market and they see why should we develop this unless there is going to be some sold. Just in the area of vacuum monitoring, Randall Phillips has put a fair sized commitment into ion gauge power supplies, for example, and so has Varian--they have a new one on the market. Well, there must be a lot of people out there measuring vacuum. But only fifteen years ago we built them because we couldn't buy them. So the market has to have some substance before business people respond to it.

STOVER: Jim brings up a good point on the size of the market base. One might ask, "what's the market for those vacuum pumps?" Quite possibly it's the IC industry that's the big market base for vacuum. So maybe for beam monitoring equipment the market base really isn't quite large enough. But where it has been large enough and with some stimulation and with big projects like the SSC you can bring that on. You know with good high voltage power supplies and fiber optics, all of that was driven by the industry but it really had to be fairly big. It needs growth. If we just have one big project over ten or fifteen years the industry will wax and wane and just disappear.

You need a consistent market. The question may be how big does the base have to be before we stimulate ongoing development effort in industry.

VAN DYCK: Part of that would have to do with the level of standardization, and I was going to ask Bob Shafer to comment on the level at which standardization might be possible. That is, you could have a manufacturer design a BPM system which consists of the strip lines and the processor modules? That sounds like an awfully specialized thing to be building because every accelerator's need is different, so it seems like you would need to standardize at some lower level.

SHAFER: It's indeed very specialized because everbody's beam line has a different diameter beam pipe, for example. In electron machines we have button electrodes and for proton machines we have strip lines, and so on. Processing frequencies are very different and therefore the front end has a very different design. The one area in the BPMs that is possibly common is in some of the processing modules themselves, like I mentioned hard limiters this afternoon. It is a very difficult job for us because they have to be individually hand made and matched but if we can get into the surface-mount technology or hybrid technology perhaps they won't have to be hand matched anymore. So there is potential application there. When big accelerators get into this and want large quantities, industry always becomes interested. I point out the SSC, which has been encouraged to go out to industry and will, I think, force commonality. Industry will certainly get involved and force commonality just because of cost reduction in large quantities built for SSC. We even saw this back when I was buying stuff for the PREP pool. A million or two dollars a year excited Walter Lecroy. We just started a band wagon there and a lot of people followed. I think it is going to be very hard to force on different laboratories to go to a common source unless the funding source demands it of us. I think that is about the only way.

VAN DYCK: Mark Ross, your laboratory brilliantly pioneered the single pass collider. Was the instrumentation for that a whole new ball game or did you find help of any kind outside of your own laboratory?

ROSS: First, the requirements for the SLC were very new. Second, we didn't know exactly what problems we were going to run into when we started to make the machine work. Most of the problems that we faced we ended up solving within the laboratory, except some we have not solved yet. That is one of the reasons why I'm here. And we found within the machine itself enough variations within the parameters involved and specifications required for particular devices that we were unable to produce standardized components except in a few cases. We found for example that we needed to measure beam sizes from a centimeter to a micron; these required very different engineering responses. Also, in contrast to Bob's story about Fermilab's construction of the Tevatron, we had very little time. We picked a particular circuit or set of circuits, systems that looked good, and replicated them as fast as possible and used them everywhere. Now we look back on it and wish we had done it slightly differently in some cases. With my colleagues here from SLAC I'm struck by the tremendous differences in the requirements of a machine like the Tevatron and the SLC. These differences are not just in the front end signals but also in the controls. For example, the timing system required to operate the SLC has to distribute signals that are stable to much better than a nanosecond throughout the three mile complex. In a big storage ring complex like Fermilab, there is no such requirement for fine timing. Our hardware has to be built to take this into account.

AUDIENCE (unidentified): Can you tell me about what's being done now with the change from main frame computers toward work stations and more distributed intelligence in the controls networks?

ROSS: Let me comment on that. Next week there is a controls workshop in Vancouver, the third one of the series. The get-together is similar in flavor to this one, so far anyway: people sit down and talk about the evolution of control systems.

AUDIENCE (unidentified): Can you comment on the impact of standardization of instrument buses like IEEE-488?

HINKSON: I can comment on that appropo the meeting in Vancouver. Steve Magure, who's responsible for the control system in ALS, will be at the Vancouver meeting and he will be representing our control system. It's a highly distributed system of five to six hundred computers and we will be dealing with RS-232, GPIB, all manner of interface of instruments to get into the control system, so we hope we have a general approach to this problem. It's something that is extremely flexible and very fast so if you're interested in what's happening in control systems maybe the Vancouver menu might be interesting to you.

AUDIENCE (unidentified): Can you say that the change now to standardized instruments has allowed you to go out to industry more readily to get the variety of instruments that you need and link them together?

HINKSON: Well, what's happening is very interesting. I do not have to go to a "one zero" guy and beg him to deal with my problems. I have to write the software to control my instruments and I'm given a format or a way of dealing with this in this control system so I can do it. At one level I have to go to these control systems people and say, "okay, now condense this stuff; put it down in EPROM in the microprocessor and make this thing run." But I'm involved from the very beginning so ultimately I'm responsible for it. What's happening here is you involve the computer people intimately; the computer people are us. We won't have this polarization that I see happening in so many places and the system integration should work much better.

We do have need to handle different interfaces--Ethernet, the GPIB, RS-232, all that sort of stuff. I think the secret here is to involve the whole staff who care about the running of the machine in the control system.

AUDIENCE (unidentified): Could somebody comment on whether the labs have certain input into standardization for these bus structures? For example, Tektronix and others recently have been coming out with this VXE bus extension of VME specifically for instrumention. Tektronix was trying to make open invitation for components to be put on that bus and therefore commercially advantageous for them to pursue. Do we have input into those manufacturers or bus structure committees which are setting up standards?

VAN DYCK: I believe that the organizing committee for this meeting decided to stay away from control system architecture because it is a bigger and harder problem which next week will consume an entire meeting. The topic we set in front of ourselves here was measuring the properties of an ensemble of moving particles.

AUDIENCE (unidentified): Not the bus structures etc. but the instrumentation which goes on it.

VAN DYCK: I'm not aware of any present mechanism for developing standards for that kind of instrumentation. That is on our agenda and is the kind of thing that we are aiming towards.

AUDIENCE (unidentified): Five years ago as we were starting stochastic cooling at Fermilab, the VME bus was the state of the art but to get Fermilab to use it was a struggle. Although Fermilab now use VME, how can you have input when you start using something five years after it was put out? Fermilab is the preminent accelerator in the world but they jump on the technology too late to have input. JULIEN

BERGOZ (in audience): What we've been talking about so far is creating a bigger daughter field to attract more manufacturers, but I think this approach is wrong. The reason I think so is that any standard which is imposed by the user starts lagging behind technology. When you have a market which is large, you should find a large company to tackle it; when the market is small, find a small company to tackle it. There are small companies in every country to do this. But in order to succeed, you have to involve your industrial partner early, not to have a closed specification by the time you submit the problem to the manufacturer. The manufacturer has had enough freedom to make a general enough design so that the manufacturer can consider your business first as an investment, then work at cost in order to recoup his investment on many more customers later.

LAUTERBACH: Yes, I think that is one of the keys, speaking from the manufacturer's point of view. If you decide that you have the time available and that you're going to design it yourself, then fine, that is what you'll wind up doing. However, if you decide in advance that you want to use your time somewhere else and you want to involve a vendor to help you with his time and his money, perhaps to do the design, then you'll find that the vendor can probably do the job. But you need to involve the vendor early and make that as a conscious decision, and maybe you'll give him some latitude that the item he designs can be used at another accelerator or somewhere else; maybe put in a few more specifications and make the market larger and perhaps you may even get a few more features at nearly the same cost. The key point that you made is that you need to involve the vendor in advance. I think that is what the SSC is trying to do and basically that is the only way you'll make it work.

VAN DYCK: Luigi Rezzonico, your area of Europe has one of the highest density of accelerators in the world. Among the big laboratories and many smaller ones are PSI/SIN, Saturne, CERN, DESY, and others. Have you found it possible to work together on accelerator instrumentation? What has your experience been in farming jobs out to industry or finding standard solutions from industry?

REZZONICO: First of all, as you know, we have an accelerator complex with a 600 MeV cyclotron for meson production. The cost of the facility is about $200M and for beam diagnostics we have invested about $2M for hardware, not counting the control system. We started this job in the late 60s when it was impossible to buy devices, so we developed our own system. Naturally, we referred to publications and we copied a lot of existing designs, making modifications to fit our requirements. Looking at standardization, I find three reasons--technical, psychological, and financial--standardization doesn't do as well as it should. From the technical point a view, alot of things don't fit your exact requirements. Maybe it's just the power or dimensions. From the psychological point of view the lab is a research lab and is intended to do interesting development work. Developing new things is much more interesting than buying. From the financial point of view, if you ask how much it costs to buy, you get a big price, so you say, "no, I can do it cheaper." Maybe this is not true, if you look at exactly what it costs, including the cost of development. And then there are requirements from our governments, for example not to buy abroad. From the standpoint of the companies, often the market is too small. For example, a year or so ago, we developed an actuator for 50-150 mm motion of slits, beam stops, or whatever, with electrical and water feedthrus and a controller. We built 30 units with a small company in Switzerland and I sold this company the right to fabricate and sell this device but they were able to sell only two pieces--incidentally, to the semi-conductor industry.

HANS GERWERS (in audience): Most arguments I heard here are quite true. There are very different requirements for diagnostic equipment. Our small company, Science International, has been selling beam diagnostic equipment to a number of laboratories and universities. We've encountered all those problems that you've mentioned here but we have been selling this equipment at a profit. Now, the strategy we follow is quite straightforward. We have a number of so-called standard products; however, since every need is different, within the frame of the standard product there is leeway do to a certain amount of custom engineering. By following that path you keep your design engineering costs to a minimum. I agree there are applications where the standard product is not able to fulfill the requirements of customers but then you have to see if the customer is willing to do new engineering.

CHRISTINE CELATA (in audience): There was something that I don't understand. It sounds very good to involve the vendor early, except if I understand the legalities you need to open your competition to a number of companies. But the earlier in the process, the harder it is to go to a field of companies and say, "please design me something that will do this and that." If you have the thing all designed it is much easier, I should think, to go out for a competitive bid.

On another topic, I don't think you build in-house just because you want the job done a certain way. There is also the fact that there is a certain group you want to keep employed at your laboratory, people whose expertise you want to maintain for other jobs. If you start farming out all your stuff then you can lose in-house expertise that you'll want later.

VAN DYCK: I think that's right. The technical manager feels like he needs to maintain the resource necessary to keep the thing going and that may mean using those people at times to build things just to keep them employed.

PRIESTER: I think you made to valid points. Most laboratories want to maintain their ability to perform their missions so therefore they want to keep those individuals that are required to do the job on their payroll. On the question about legality, you're right, you cannot go to one particular company and say, "I want you to build this," because competition in contracting is required. But early on, you can write a specification and go out on bid.

LAUTERBACH: The one thing you should get is more dialog between the accelerator people and the industry people. I don't see anything wrong with working with industry in advance. You do not have to wait to the point where you need to buy something out of a catalog. You know often you can get good ideas from industry as opposed to going to them with your ideas and it can help everyone out.

AUDIENCE (unidentified): That brings up the point of exchange of intellectual property in dealing with companies. In the national laboratory system, our engineering ideas go into the public domain, whereas companies are not necessarily under that requirement. Maybe when trying to set up a rapport with a company, we are the ones with the initial idea, so there is a flow of information from us to them but there's not necessarily the same flow of expertise and information back from the people in the company. It seems like it would be helpful if there were either better rewards for patent and intellectual property within the national science system or more openness on the part of the companies.

VAN DYCK: Perhaps this would be an appropriate moment to ask Don Priester to comment upon the SBIR funding cycle and what is the legal status of the intellectual property generated by a company when it receive an SBIR grant.

PRIESTER: For those who are not familiar, SBIR program is the small business innovative research program instituted by Congress about 1985. Congress was worried about the level of innovation coming from small businesses and that a lot of the things developed by the federal government were not being commercialized. What they wanted was to involve these small companies in research and to produce products which could be commercialized. They put this program in place in 1985 to encourage small businesses to come in to DOE with their innovative ideas, "innovative" being open to a lot of interpretation. Then they offered funding under Phase II to prove whether these ideas were feasible for production. A third phase was to follow that whereby these things would go into commercialization. DOE found the program to be very successful, at least in the area I came from. Under Phase I, a company gets $50K just to show that they can do something and that led to maybe half of the awardees moving into Phase II where they got up to $500K over a couple of years. Now the ideas that are developed under the SBIR legally belong to the company, according to the way the program was set up. However, in the last couple of years some of the programs within DOE haven't worked as well.

Let me back up: I do not represent the SBIR program, but I have been involved with them for about five years as technical topic manager and as a technical project manager. The technical topic manager is the person who sets up the topics for proposals and aids in evaluating these proposals. The technical project manager is the person who follows it once it gets awarded. Now as a TTM, I found that a lot of the ideas that people come in with run into problems mainly because they didn't contact the laboratories beforehand. I think that is very important. There are two bosses: first it is DOE and then it is the laboratory.

SHAFER: I have evaluated SBIR proposals sent to me by DOE and I find that these small companies sometimes are really not as well informed as they should be before they start proposing these things. I think the onus is on them to go out and understand better what the industry needs or what the national labs need before they make these proposals. Some of the proposals don't really show awareness of what's being done.

PRIESTER: I have to admit that many of the ideas presented have not been very well thought out. You have to keep in mind that these are small companies. What you have to look at is whether the proposed idea is good and evaluate the idea on its technical merit. Maybe you try to underplay the omissions. The SBIR solicitation is published, I think in CBD, but there is also a solicitation which is mailed to everybody who applies or who we have on a particular list. We try to put a list together of everybody we can possibly think of and we send a solicitation to them. It includes details on the particular subject. There are about thirty topics and the topics have a bibliography on related ideas. There's a telephone number for the SBIR office where they can refer you to the individuals who handle that particular topic and who will normally point you to the laboratory you need to contact.

[**QUESTION** from the audience on reliability and support on products built by small companies...]

PRIESTER: These companies are treated just like any other company. Whatever warranty they give on the particular product is what you get; I'm not sure that you can get a full 100% warranty on anything in this country today.

LAUTERBACH: The shoe sort of fits on both feet here. We've run across laboratories where sometimes for an in-house design, maybe the designer's still there but the guy who built them left, or maybe they ran out of spare parts and the guy who used to sell them the critical diode doesn't make it anymore. You can be in the same boat either way. Generally what I have found, and I have worked on both sides, in the field and in the industry, is that there is a higher level expected for documentation and support from industry than there is from in-house design. With the in-house design you expect to have the guy there if you need some help. Industry is rightly expected to produce more documentation to support repairs for five or seven years. In the state of New York you have to support your equipment for at least seven years. You do not necessarily have that protection on an in-house design. If a company goes bankrupt, of course, then perhaps you've lost your protection. I would say just choose your vendor carefully the same way you would choose carefully the people who would work in-house. Generally I think the level of support, documentation, calibration procedures, training courses, and spare parts availability tend to be better from something that is commercially available, especially if it's sold to several laboratories.

HUBBARD: The SLAC scanner-processor is a good example. It is effectively an in-house-built FASTBUS computer from numerous different laboratories that did a little bit of a development on it and left no documentation trail whatsoever. Over here in the physics department, there is a pile of them ten or twenty high that have just been completed and they cannot get to work and they've given up. That is because of the total lack of documentation.

JACKSON: I'm the group leader of the instrumentaion group at Fermilab. I have had a lot less trouble from the stuff we bought outside the Lab than the stuff we had built inside the Lab. But there's a problem with the guy who signs the check. I cannot buy most of the equipment I need because of the capital equipment restriction. Capital equipment is a separate fund from the rest of the money and we are only allowed a small amount of equipment money because we are low priority compared to the detectors and the big ticket items. Some years we're allowed to buy only $40K worth of equipment.

VAN DYCK: This is another kind of management problem that the national labs have and I guess we expect managers to hear this and respond. I would like to ask Robert Odum, as the representative of Charles Evans and Associates, what his experience has been with the SBIR program.

ODUM: We've been reasonably successful getting funding through SBIR for Phase I and Phase II, and ultimate commercialization. A lot of our projects today have been things that we have identified a need for in our particular business, which is materials analysis. We have begun to expand in areas where we have less assurance of the potential business there, one area actually being in detectors for particle beams. In that particular area we've actually surveyed interest by mailing out to three hundred people some preliminary results we've gotten on a Phase I project funded by DOE; we got just a few responses. One thing that I would like to stress especially in this area is that the managers in the labs should tell the people at DOE what they need, what they'd like to see. Maybe DOE needs to do a survey on what could be standardized on the accelerators. The tiny response we got to our mailing indicates that either we're off base or it's just the wrong product. One other thing we've observed is that it's about a three to five year turnaround in getting research money to develop a product and actually getting a product out in the market place. From the SBIR standpoint, it is not something where I can just call up the vendor and say I need a product. It is more in research area, as Don said, something innovative.

AUDIENCE (unidentified): I'd like to make two points: first, capital equipment is a lump sum and there seems to be a lot more reluctance to just plump down that money to get what you need. The attitude is, well, we can probably build that ourselves in-house it will probably be less expensive because the chip set only costs $25. Then you whittle away at the engineering time repairs, and documentation time until the end cost is higher. There seems to be a problem with planning ahead. Also, commercial products tend to be reused whereas modules built for specific applications do not get recycled. My second point on procurement relates to large projects like the SSC. Is there going to be a move to bring in multiple vendors? I think some military projects do this where the lowest bidder gets the majority of the project for a time; then there is a reevaluation period and the laboratory can switch to a more acceptable vendor, who might be the second-lowest bidder, or otherwise depending upon previous performance.

PRIESTER: As far as the SSC is concerned, there are a number of options that are being explored at this point; that means multiple vendors and everything else. There is a lot of interest in having more than one vendor involved in these procurements. I think people realize that when you get a very large program that you do not want to have it dependent upon one vendor. These issues are still under evaluation at this point.

MELCHER: My name is Jerry Melcher; I'm now with Analytek and formerly with DSP. From my viewpoint in the commercial marketplace, the pool of money that is available in high-energy physics is not growing very fast. I know from the point of DSP Technology and Lecroy that if we had not gone out to find applications in the commercial marketplace, we would have not seen any type of growth. There definitely has to be a commitment by managers in the labs to start restricting the number of units that are built in-house, say at a maximum of ten units. After that, you license the unit to outside vendors and you provide engineering data to replicate the unit, provide a license to supply to other DOE facilities, and allow design improvements for competitive reasons. One unit in particular that comes to mind is the "2032," the 32 channel DVM designed by SLAC that wasn't put out successfully into commercialization. I think the last time Fermilab went around for bids on that unit, Kinetics picked it up. It was originally selling for about $2 400 a unit, but I think now it is somewhere around $1 200 to $1 400. You know, more competition like that does foster a lower price. The multiple vendor bidding system does give the labs a break. Gerry Jackson made the good point about limited capital expenditure budgets; that has to be addressed as a management problem. You're not going to get anybody who's going to come into this business and supply it unless you're waving around big dollar items like the SSC. The only reason Tektronix, HP, Racal-Vadic, Colorado Data Systems, all these people are paying attention to SSC is because someone is waving a billion dollars, saying "we are going to spend a billion dollars on controls and instrumentation."

VAN DYCK: The same point or question has arisen repeatedly this evening. Let me try this on Mike Lauterbach: virtually every experiment on every accelerator takes its data through NIM and CAMAC modules. Why doesn't the beam for that experiment get tuned through standardized NIM or CAMAC modules? Why isn't the market comparable? The market for NIM and CAMAC did not start with a billion dollars twenty years ago, it was nowhere near that big, but yet those standards did evolve. Why hasn't that happened yet in our business?

LAUTERBACH: That's a good question. Lecroy started out 25 years ago making equipment for high energy physics and most of our new designs are based on what we hear people telling us they want for the future. In the case of the detector readout, whether that is a detector in a beam line measuring the profile or a detector looking at what scatters out of proton on proton, we try to listen to what the market tells us which is really what individual people like you guys tell us. We even tried in a few cases, like on secondary mission monitors, to design a product in advance which we thought the accelerator industry would want to buy. In that particular case, we designed something called PICOS which was a wire readout system. We thought that every accelerator around the world would probably want to use it but it didn't really work out. Maybe you guys in the accelerators can tell me more about that why they decided that they could make something a little bit different that would be better for their machine, or why they wanted to work on something themselves. Maybe they didn't even wind up with something better but they made it and they decided they wanted to use it. A little too much I think of "we would rather do it ourselves." In your general experiment you have maybe ten thousand wires or these days even an order of maginitude more. I suppose the number of wires doing beam profiles is much less than the number of wires that are in the experiments. But that's is still an awful lot of wires and it's not clear to me why the accelerator physicists do not like to use the same kind of wire readouts as the high energy experimental physicists. It is a bit of a puzzle to us. We have a lot of questions right now about what would accelerator people like to have. It's a bit frustrating to us. We visit places like CEBAF trying to figure out what each detector group might want, whether that's an accelerator or experimental detector, and we have questions like what kind of ADCs do you want, do you want ten bits or twelve bits, separate gates or common gates, CAMAC or VME, if VME, what size, etc. We're willing to do all these things if you want or if we even have a reasonable guess for risking the investment, but in the cases where we've tried, it just hasn't really worked out all that well. For experiments, the physicists are forced to get their electronics through the pool and the pool makes sure that they have good things, but when we build an accelerator, the pool doesn't exist, so the guys who build the accelerator decide what they want. There is no forcing function; the time is there, the money is there, so why not do it ourselves. I think that gets to be expensive after awhile. You have all these accelerators around the world and they all more or less need the same sort of measurements, maybe on different scales, but nevertheless they are all measuring beam profiles and beam current.

ED BEADLE (in audience): I'm Ed Beadle, involved in AGS instrumentation. Even if we agree that we want similar things, how can we disseminate that information amongst ourselves? How can we get reference material at Brookhaven to see what somebody at Los Alamos has already developed?

VAN DYCK: What is the mechanism for developing the information about what we have and perhaps working our way towards a standard? I think that is your question, and one of the reasons why we had this workshop.

HINKSON: We've talked about some of the things we might do in this area. One thing might be a computer bulletin board. Fairly frequent meetings of people who are interested in this area is a good way to exchange information.

AUDIENCE (unidentified): Or use a video telecom. DOE has funded a national software center and I believe Los Alamos provides a central location for codes related to accelerator technology. Maybe there could be some variant on that for accelerator instrumentation.

SHAFER: Every laboratory or I guess every individual has a certain amount of ego in developing something and he'd like very much to see another lab use it, but sooner or later somebody has to come along and say what particular module or standard has to exist. Maybe it is not the best way for a lab to advertise its modules and try to get other labs to use them. I would like to point out for example our early history in NIM. As I recall from the 60s when I was in Berkeley, this was when transistors first came out, there were some modules that looked vaguely like NIM modules that really weren't: Chronetics had their funny module, University of Illinois had a very interesting logic scheme that used open collectors and was eventually adopted as NIM but I do not think they had a NIM module. There was a guy back at NBS by the name of Lou Costrell who was able to consolidate the interest of all the people that were using this kind of logic to standardize on module size and signal levels. Then modules were all built to these specifications but there is no limit on the functions they can do. Industry started building modules to these standards. They became quite popular beginning in 1963-64. Now exactly what the leverage Lou Costrell had I do not know, but he was very effective in doing that.

VAN DYCK: Once the product was there, it became very popular, which prompts me to ask Rex Hubbard whether he feels the need or the product should come first?

HUBBARD: Incidentally, let me say that NIM is not a product, it's a packaging scheme. The packaging has no data handling capabilities but it sets electrical and physical standards. And it made eminent sense to everyone not to build in a different shape box. If the question is, did the need come before the standard, then yes, by a long shot. There were all these different modules being made when the standard came along and the thing crystallized all of a sudden. Now in the case of CAMAC, the need also came before the standard. Having listened to a couple of lectures here today, I really do not think that this field is in any sense ready for a form of standardization. Those standards were basically written by the users or the potential users, not in any sense by the manufacturers, unlike all the other standards you see today like the bus standards, like VME or UNIBUS, which are all essentially driven by a given manufacturer, who then gets himself a little consortium so it seems like it is a little more general. These standards were very effectively frozen at a certain point in time and I doubt that your technology is ready for somebody to say, "alright, it may not be perfect, but we are going to freeze it right here and now and from now on for the next 10 or 15 years we are going to use equipment that works exactly like this." You are not ready for that.

LAUTERBACH: I don't think it has to be that dark. We have our new 1990 catalog here, and you know, there is NIM in it, there is CAMAC, FASTBUS, VME, there is interfaces to PCs, and interfaces to Microvaxes, there is just about everything in it. But within any given segment, if there is a new flash ADC that comes out or you need more bits or higher speed, you can put it into any format as long as there is some way to get the data out. You can always take advantage of new technology. Only maybe when the data rate gets too high, you need a new standard like the FASTBUS. If industry knows what the market is going to want, we will try to do it. The comment that was made by Jim Hinkson that LBL invented physics; I guess I'd say that physics invented Lecroy. Lecroy was strictly a high energy physics company for more that 15 years and they grew from zero to a decent size small business just from high energy physics. Now that high energy physics money world-wide is relatively flat, we have taken some of the technology they used to measure small fast signals and we put it into different kinds of instruments, like our new fast scopes. But the roots of the company are still in high energy physics.

AUDIENCE (unidentified): I have recently come into the high energy physics community from the outside, and I have some impressions that might bear on the lack of standardization. Number 1, the entire high energy physics world is microscopic compared to what's required to generate a real standard. An example of that is CAMAC; CAMAC is familiar to everybody here but 95% of the electronics people in the entire world have never heard of it, much less have ever used it. The same is going to be true I'm sure of anything that originates in the high energy physics community. What I'm getting at is if we want to find real and meaningful standards which will lower costs, we have to look outside the high energy physics community. Point number two is that it seems to be very difficult to get across to people in the community that there are standards out there that are useful. To see large scale reductions in the cost of doing high energy physics, we're going to have to look outside the physics community for applicable solutions and we're going to have to be willing to make adjustments to specifications to allow external solutions to succeed.

LAUTERBACH: Actually, it is possible even within high energy physics to have enough momentum to generate a standard: the IEEE FASTBUS standard is a good example; it is strictly high energy physics. Nonetheless, what you said is true: if it were possible to have products that were used by more than just high energy physics, there could be lower costs.

UNIDENTIFIED: An example that was mentioned is vacuum.

LAUTERBACH: Sure. Costs were not driven down in the high energy physics market but from other people. If we could get to the point where many accelerators could use the same thing, that would be nice, but if you could use the same thing that some larger businesses used, that would help you even more. One possibility I guess is

VME. There are a lot of different kinds of businesses that are using a host of processors and other instruments that fit into the VME type of architecture. If you can use a standard that is shared among various laboratories or industries, you get a lower cost. Right now I think we'd be happy if more than just one acclerator used the same kind of electronics!

BERGOZ: Yes. I think FASTBUS is a very good example. I had to manufacture some FASTBUS boards and found that manufacturing con-straints were not taken into consideration when FASTBUS was designed. It is a format which we just cannot manufacture economically. Just consider one thing: the size of this board, 400 mm by 400 mm, requires a soldering machine that costs about a half million dollars. That's not such a large investment for industry, but who is interested in a machine worth half a million dollars to solder twenty boards? Yet this is about the size of any FASTBUS order. It is nonsense.

AUDIENCE (unidentified): Let me point out that CAMAC is also an IEEE standard, and yet there is so little volume in CAMAC that a very low performance, practically primitive, microprocessor board costs 6 times what a more powerful board would cost in STD and three times what it would cost in VME.

VAN DYCK: We're nearing the end of our alloted time, but the CEBAF representative up here has not yet told us what we might do for him--what could the established laboratories do to contribute to the CEBAF instrumentation effort.

ROSSMANITH: CEBAF is a very small laboratory and the I & C group is even smaller so we depend on industry very much and on help from the outside. We are buying as much as possible from industry; for instance, our current monitor system is built in France which is a little bit complicated and expensive. Our actuators for intercepting monitors are built by an American firm at least to 80%--the rest is done by a local firm. We have collaborations with many other laboratories concerning software development, for instance, NIST and Berkeley; even DESY helps us a little bit. So we really depend on other laboratories and industry to develop our instrumentation system. Our beam position monitor is being prototyped at CEBAF and we expect to bid it outside. We have limited workshop capacity at CEBAF so we cannot build it in-house anyway. We have a consulting firm for electronics so that the monitor electronics is about 75% developed outside CEBAF; we are just looking over their shoulders and making the tests. We definitely tried to contact other laboratories, for instance Los Alamos, but they were very busy, which I do not mean as a criticism. We tried this with various other laboratories and I think we really hope that we will get more help from the outside and some laboratories can help us to develop or advise us in our instrumentation system. I really hope that we find interest here in this auditorium.

AUDIENCE (unidentified): Isn't there some kind of list of companies who do accelerator instrumentation?

AUDIENCE (unidentified): The NBS developed a list of suppliers for NIM, CAMAC, etc. a number of years ago. There are user brochures put out by Fermilab for FASTBUS, maybe also for CAMAC. The other thing I would like to point out is that high energy physics has one of the best communication systems in the world with BITNET. Why not give commercial suppliers access to it? We do not want to advertise on it but we certainly like to be able to be out there to answer questions.

VAN DYCK: The advertisers in Physics Today form a kind of list.

HUBBARD: There is a publication from CERN that used to come out quite regularly that listed at least all the CAMAC modules available from any country in the world.

VAN DYCK: Do the panelists think that the time is ripe to start formulations of standards for accelerator instrumentation? For example, Bob Shafer, in rf processing for beam position monitors?

SHAFER: There are certain areas of work for the SSC which just by the very quantity might provide some incentive for use at other laboratories. I'll point specifically at RHIC which is going to be built here in the very near future, also ALS and APS which are two electron synchrotron light sources that are going to be built soon. But I suspect the only driving force we are going to see in the near future is the SSC.

ROSSMANITH: I think we have such devices, for instance, the famous current monitors which were described this morning and are used by many laboratories. We have at least one firm which is building them on an economic basis. The position monitor is a little bit more difficult. Profile monitors, wire scanners, and so on should be standardized. I think that everybody is using similar design. There are firms that are already offering such intercepting monitors but maybe the requirements and the offerings are somewhat different, so we have to work on this.

ODUM: It is not really clear to me whether there could be viable standards at this time, although I'm not very familiar with this field. In business, you either have volume sales or you make a few that cost a lot. In our company I know that our selling price typically tends to be three times the development cost so if we can see that kind of a return on an investment then we'll probably go for it even if it's a small item, but we are a small business.

LAUTERBACH: I think with the number of projects being built now, there clearly has to be some benefit in having interchangable parts, even just for the physicist who works on one accelerator and then at another. Also, there's the cost savings up front if you can build one design and use it in two or three different places--the learning curves, parts and labor savings, all flow together and they help everybody out. It seems clear to me that there is a lot to be gained through standardization and now is probably the best time. The machines are bigger now and need more of everything to run them so it becomes more commercially viable.

REZZONICO: Internally we have standardized a lot of things--profile monitors, actuators, and beam current readout.

WEBER: I guess I cannot point to any particular thing at Fermilab where we could say we are ready to standardize in cooperation with everyone else but maybe we've taken the first step by all agreeing to talk about it. I think we owe some thanks to the people at Brookhaven and Dick Witkover and Gerry Bennet for organizing this. We also owe thanks to the commercial companies for showing up, letting us know what their capabilities are, and how we might better work with them.

HINKSON: It's not clear to me. I have $1.5M committed to 150 beam position monitors but it has taken me $300K to understand the problem. Now I have to build a good deal more and I'm wondering if there's anyone out there who will build in a Eurocard crate, interface to our control system, and give me RF measurements to a part in a 1 000 with a 10 MHz bandwith, and do it for $1000K? I wonder about this. I think we are small potatoes and have very specialized needs. In some areas of course we take advantage of standardization. We may have a common format, a common crate, common power supplies. But our needs are so terribly unique I cannot imagine why industry would want to be involved, like in the area of beam position measurement where it gets really tricky. With the SSC, however, that is a different ballgame--10 000 magnets in the ring! But for us at ALS, a $100M machine, we've got some pretty sophisticated needs. I do not know how we can involve industry any more than we have.

ROSS: The SLC posed a lot of problems that were completely new and I think next generation linear collider will do the same thing. We'll need new ways to measure the sizes of nanometer beams and I do not think that there is anybody out there who claims that their profile monitor technique can be extrapolated to that requirement. However, that is not the whole story. There are many mundane things, like profile monitors looking at fairly big beams of order 100 microns or so. We were able to buy video processing hardware and various commercial products for that type of thing and fit it right in with a little bit of shoe horning. On position monitors I guess I agree with Bob; it seems hard to expect something even approaching standardized which is going to be able to fit most of our needs. In the SLC, there are two thousand position monitors. As far as I know, this is one of the largest installations around and they are somewhat specialized. Maybe we could have involved industry in the design and production of these things, or we could have had a hybrid manufactured for the most sensitive parts of the circuit. But at the time, we didn't feel we had the time to do that. What I find from the discussion here tonight is that there are services and products that I didn't know about. I read the CERN Courier and Physics Today. I read the trade newspapers and stuff in the mail but I didn't know about companies making wire scanners. The people who are building things like that should be a little bit more aggressive in letting potential buyers know about their products.

VAN DYCK: Richard, from what you heard tonight, do you think there is enough commonality of interest so that at least two or three or four labs should be getting together?

WITKOVER: Let me start by saying that I agree with Bob Shafer that the large projects are the ones that are going to drive us to some kind of standardization. SSC, just by its overwhelming size and price tag, will create a standard just the way IBM did, and certainly it would be to our advantage to work with them in any way we can to make that standard something that we can use as well. I hope we can talk to them about our requirements and perhaps fit them together. I think the committee here brought home that standardization is a huge task, a very difficult thing, with many levels of problems involved. To try to begin making a decision here tonight is certainly not realistic. The best that we can say is that we are beginning to talk about it and we certainly perceive a need at least in some areas. Perhaps the first thing we can do is maybe look at a common packaging so that electronics modules that are developed in one laboratory will at least fit in the same crate at other laboratories. Many of the things do not require a bus structure, they are analog signals, and so the complexity of having an IEEE standard developed maybe do not exist. If we start with something finite, such as the packaging, it would be a major accomplishment in itself. It has been brought up that we'll have to continue discussions, and I would like to see something established on that line for the future.

AUDIENCE (unidentified): I'll be traveling around the country for the next ten years or so helping to put the SSC together and I would like to be able to call on people to share with me the engineering wisdom from the past. I know that a lot of scientists and engineers get a kick out of seeing somebody else use their work and we'll certainly be able to do that with the SSC. From the discussion here, it sounds like people would be very open.

AUDIENCE (unidentified): I see a lot of businesses represented here. Would it be appropriate to have each of them tell us briefly what their areas of expertise are?

VAN DYCK: Sure. Let's go around the room.

ODUM: Charles Evans is primarily a materials analysis company though we've been involved with imaging detectors for low energy ion beams. We are presently in Phase

II of a DOE SBIR project for what we call a nondestructive particle beam detector. The basic concept is that we take a particle beam through a thin carbon foil and make either secondary electrons or secondary ions which are stigmatically imaged onto a two dimensional detector, either a micro channel plate. or a fluorescent screen. We are looking for applications within the high energy physics community for this type of detector. One can envision putting two of these in series and have a beam emittance monitoring device, for example. We are actually using this in collaboration with some people at Sandia-Livermore in a proton beam tomography unit, using it as a two dimensional imaging detector.

BURKE: My name is Kevin Burke and I'm with Advanced Technology Laboratories. We are a combination design and manufacturing firm and provide one of a kind up to small volume manufacturing of electronic or electro-mechanical systems. We provide electronic design, packaging design and manufacturing services.

RADWAY: I'm Tim Radway; I work with Jorway Corporation. Basically we manufacture CAMAC equipment, general purpose I/O gear and interfaces between CAMAC and a number of different processors. I would like to try to reiterate what has been said a number of times before: I think the word is involvement. Many of us in industry would like to be involved in your problems; our involvement primarily is because we hope eventually to build something we can manufacture for profit. It is a two way street. We have to be involved in pursuing people in the laboratories to find out their problems, and you in the laboratories have to give your time freely to explaining what your problems are. By talking to a number of you we may find commonality in your problems so that we can visualize a product that might serve more than one customer; maybe this technology might even create products that we can sell in industry. What we're looking for is volume and the only way that we are going to find it is by talking to you people, finding out what your problems are and seeing whether there is some kind of design that makes sense for us to manufacture. The statement that industry has been saying to you is that we need involvement.

BERGOZ: My name is Julian Bergoz and I have been working with my small French company two kilometers from CERN on the French side of the border. We have been working for the past four years with Klaus Unser on a collaboration contract for beam instrumentation. We developed under this contract a Parametric Current Transformer (PCT), which is a beam current monitor covering the frequency range from dc to 100 kHz. At present its upper full scale range is in the range of 2-10 amps and the resolution as seen in the one second integration window goes down to 300 nanoamps for the standard out-of-catalog instruments. We are working now on a collaboration for GANIL, together with Mr. Unser, to push the resolution down to about 50 nanoamps. We have in the process of this collabration developed other PCTs with about 35 units sold to quite a few accelerators under construction or being upgraded now, including of course LEP and the PS at CERN. They have installed quite a few units on HERA, also on the TSR accelerator in Heidelberg. In Italy we went to the synchrotron in Trieste and to Adone in Frascatti. In France there was one being manufactured now that will go to GANIL. In the U.S. Mr. Rossmanith had mentioned that we have delivered a couple of units to him; I think one of those units will go to Fermilab on loan for a period of time. We have three units on order for Brookhaven National Lab and a unit for the Loma Linda Medical Accelerator. So I think that pretty much this popular unit seems to be covering a very wide range of needs and I would not mind it being the defacto standard even though I seem to be against standards. It is a one way street but for me I'm quite happy with it. We have been working on other instruments like the active- passive transformers that were talked about this morning, and also very fast basic transformers with a rise time of better than a nanosecond. Through the collaborations with CERN we have also been developing a design which belongs to them for a bus coupler that is an interface between the Macintosh II and VME or CAMAC. This interface allows mapping of up to 8 VME or CAMAC crates into the address space of Mac microprocessor. This allows direct access to any VME or CAMAC equipment by the Macintosh without going through subroutines or other interfaces. There's direct addressing of CAMAC or VME through this and it is a pretty fast interface using a parallel dataway through twisted pairs with up to 100 meter cables. The transmission rate between the Macintosh and VME crate is one microsecond for 16 bits. Thank you for giving me the opportunity to talk about it.

GERWERS: My name is Hans Gerwers; I'm with Scientific International. We are a straight-forward sales marketing company for a number of high tech products. As far as beam line diagnostic equipment is concerned, our designs are primarily based on GSI designs (the accelerator facility in West Germany). Among our products we offer Faraday cups (cooled or uncooled), beam profile monitors, rotating wire scanners, beam stops, emmittance measuring systems, capacitive pickup probes, actuating mechanisms using either air or stepping motors We offer fast-acting flapper valves for the protection of beam lines and a number of so-called standard systems which can be customized to some degree and maintained and serviced by Tektronix.

JOERGER: My name is Fred Joerger and I represent Joerger Enterprises. We've been in business about 17 years. We make about 60-70 CAMAC modules. They cover the range--ADCs, DACs, counters, registers, stepping motor controllers, and drivers. We're just finishing a Phase II SBIR project for transient recorders; we've developed five transient recorders, the fastest is 200 MHz with 256K memory in a double-wide module and is reasonably inexpensive. Another is a dual 25 MHz version with 64K memory per channel; also we have a 25 MHz VME transient recorder with 128K of memory, and a 16 bit, 32 channel scanning ADC and transient recorder with 64K of memory. These products, just being finished now, are sponsored by the government.

MELCHER: My name is Jerry Melcher. I'm with Analytek Corporation and formerly with DSP. Analytek is a new start-up under Ray Larsen and Bob Meldon formerly at SLAC, funded partially with venture capital and with some input from Tektronix. We manufacture VME-bus-standard high-speed multichannel waveform acquisition systems with up to 2 gigasamples per second, up to 12 slots in a 19 inch rack. It's very inexpensive, with some of the best resolution in the industry, 10 1/2 bits of dynamic range. Tektronix services our products worldwide.

ULENAS: I'm James Ulenas, the head of Vetra Systems Corporation. We're a custom engineering company here on Long Island, where we can manufacture in small to reasonable quantities. I would like to pass on some thoughts that I had on this industry. One of the questions that I hadn't heard anybody ask is why standardization--to reduce cost, to reduce the effort in learning if you fly the same thing from generation to generation? Standardization is not neccessarily the panacea for cost. Just this morning in the Times, McGraw Hill announced that it would customize college text books down to quantities of ten per professor. So volume manufacturing is not the only one way for us to get a reasonable return with reasonable profits. There are other ways; I think that you in the community should ask how can you involve industry to help you. Standardization per se is not necessarily the answer. Standardization on screws is a good thing. Standardizing in high tech can be a problem. I remind you of PL-1! Standardizing high technology locks you into a particular state of the art which is obsolete by definition.

VAN DYCK: Let's close this session with thanks again to all the participants, especially the business representatives for whom time is money.

Accelerator Instrumentation Workshop
(Closing remarks and organization of next workshop)

Greg Stover

I. Summary of closing remarks, Organization of next Workshop:

The closing session of the Accelerator Instrumentation workshop was convened to present a tabulated summary of the workshop evaluation questionnaires and to solicit comments and ideas on the organization and format for the next workshop. The following text is a restatement of the results obtained from the questionnaires and a distillation of the ideas and comments obtained from the conferees via the questionnaire and the closing session discussion that followed.

The questionnaire, voluntarily filled out and returned by the conferees earlier in the week, was composed of nine questions; six multiple choice and three essay type. From over 100 conferees attending the workshop there were 46 individuals who returned questionnaire. Among the multiple choice questions a resounding 100% of the respondents agreed that future workshops should be organized. Approximately 54% preferred a one year interval between meeting dates followed by 38% who selected the two year time span. A workshop duration of three days was selected by 64% with another 30% favoring four days. A strong majority ,78%, preferred that subsequent workshops should cover a very broad array of beam monitoring topics with 20% desiring a more limited menu. On the question of format, 82%, preferred a combination of invited and contributed papers. The preference for the style of the workshop was more evenly mixed with 58% favoring a combination of oral and poster sessions and 40% opting for oral sessions only.

On the essay question "What aspects did you like best?" there were a number of consensus points among the respondents. Of these, meeting other people in the field, the cross-fertilization of ideas, bringing in industry representatives and open discussions were the most commonly cited benefits. The desire for face to face interaction between the participants was reenforced by the post discussion comment that future workshops allot time for small informal discussion groups and possibly poster sessions. The respondents also

Accelerator   Instrumentation   Workshop
(Closing remarks and organization of next workshop)

strongly affirmed their desire for the continued use of the pre-tutorial outlines and comprehensive reference lists.

Of those aspects of the workshop deemed to be "not very worthwhile" or reinterpreted by some to mean "could be improved" there were several propositions made by the respondents. Though all the participants acknowledged that the presentation of mathematical theory during the tutorials was important many people expressed the desire for more data about present and future implementations of the device being discussed. Specifically, people felt that more data should be included on what other conferees were currently doing in their respective laboratories i.e. descriptions of actual beam monitoring systems, their operating characteristics, performance data, and the methodology of construction.

A second aspect that was viewed in need of improvement was the commonly expressed desire for more than one speaker during the long view tutorial sessions. And, as suggested in the discussion, possibly three or more speakers over a three hour period, with one invited and several contributed papers, might provide greater diversity and flexibility.

C   Conclusions of survey

The conferees definitely indicated a desire that future workshops be more hardware oriented and that dialogs with equipment manufacturers be continued and possibly enhanced with product demonstrations. In keeping with this "engineering" preference it was suggested that future workshops address topics related to new projects coming on-line such as the Relativistic Heavy Ion Collider (R.H.I.C.) and the Super Conducting Super Collider (S.S.C.), and that an on-going standards committee be set-up to advise the national accelerator laboratories and commercial equipment manufacturers.

In the area of format and presentation the conferees did split their preferences between those relatively new to the field who

Accelerator Instrumentation Workshop
(Closing remarks and organization of next workshop)

desired a more general tutorial style and those with more specific
interests and experience who would prefer to attend only small
informal discussion groups. It was suggested in the discussion that a
format similar to the Nuclear Science Symposium be adopted;   that is
to begin the first day or two of the workshop with tutorials on a
limited number of topics that would then be followed by small
specialized discussion groups and possibly poster sessions in the later
half of the program thus allowing people with differing interests and
levels of knowledge to choose a program that would fit their busy
schedules. Most of the conferees agreed that the future workshops
should be kept relatively small, around or less than 100 people.

II.   Acknowledgements

The members of the organizing and selection committee would
like to thank the speakers for their time and effort in producing and
presenting their respective papers and all the participants for their
thoughtful and considered participation throughout the workshop.

BROOKHAVEN NATIONAL LABORATORY
Upton, New York 11973

List of Attendees

Richard L. Abbott
Strategic Defense Command
Huntsville, Alabama 35807

Jeffrey A. Arthur
MS 308
Fermilab
P.O. Box 500
Batavia, IL 60510

Patrick Ausset
CEN-Saclay
Laboratoire National Saturne (Service
  Machine)
91191 Gif sur Yvette, Cedex
FRANCE

Alan Band
MS H838
Los Alamos National Laboratory
P.O. Box 1663
Los Alamos, NM 87545

Walter C. Barry
CEBAF
12000 Jefferson Avenue
Newport News, VA 23606

Edward R. Beadle
AGS Department, Bldg. 911B
Brookhaven National Laboratory
Upton, NY 11973

Gerald W. Bennett*
AGS Department, Bldg. 911B
Brookhaven National Laboratory
Upton, NY 11973

Julien Bergoz
01170 Crozet
FRANCE

Richard W. Biscardi
NSLS, Bldg. 725C
Brookhaven National Laboratory
Upton, NY 11973

John W. Bittner
NSLS, Bldg. 725B
Brookhaven National Laboratory
Upton, NY 11973

Richard A. Blue
Cyclotron Laboratory
Michigan State University
East Lansing, MI 48824

Joseph M. Brennan
AGS Department, Bldg. 911B
Brookhaven National Laboratory
Upton, NY 11973

David Brown
MS H838
Los Alamos National Laboratory
P.O. Box 1663
Los Alamos, NM 87545

212

John Byrd
Wilson Laboratory
Cornell University
Ithaca, NY 14853

V. Castillo
AGS Department, Bldg. 911B
Brookhaven National Laboratory
Upton, NY 11973

Christine Celata
MS 51-208
Lawrence Berkeley Laboratory
One Cyclotron Road
Berkeley, CA 94720

Yu-Chiu Chao
Bin 12
Stanford Linear Accelerator Center
P.O. Box 4349
Stanford, CA 94309

C.L. Chen
Bldg. 902C
Brookhaven National Laboratory
Upton, NY 11973

Youngjoo Chung
Bldg. 360, C229
Argonne National Laboratory
9700 S. Cass Avenue
Argonne, IL 60439

D. Ciardullo
AGS Department, Bldg. 911B
Brookhaven National Laboratory
Upton, NY 11973

Johannes Claus
Bldg. 1005
Brookhaven National Laboratory
Upton, NY 11973

Michael David Cole
M/S A01-26
Grumman Corporation
Bethpage, NY 11714

Joe Curtiss
Nuclear Energy Department,
    Bldg. 902C
Brookhaven National Laboratory
Upton, NY 11973

Roy I. Cutler
National Institute of Standards
    and Technology
Bldg. 245/B119
Gaithersburg, MD 20899

James D. Deaton
US Army Strategic Defense Command
Huntsville, Alabama 35807

James Delrossi
Vetra Systems Corporation
1670 Old Country Road
Plainview, NY 11803

Brian Fellenz
MS 308
Fermilab
P.O. Box 500
Batavia, IL 60510

Charles L. Fink
Bldg. 207
Argonne National Laboratory
9700 S. Cass Avenue
Argonne, IL 60439

John N. Galayda
NSLS, Bldg. 725B
Brookhaven National Laboratory
Upton, NY 11973

Hans D. Gerwers
Scientific International
141 Snowden Lane
Princeton, NJ

John Douglas Gilpatrick
MS H808
Los Alamos National Laboratory
P.O. Box 1663
Los Alamos, NM 87545

James A. Hinkson*
Lawrence Berkeley Laboratory
One Cyclotron Road
Berkeley, CA 94720

Susan Hobbie
Nuclear Energy Department,
   Bldg. 902C
Brookhaven National Laboratory
Upton, NY 11973

Curt Hovater
CEBAF
12000 Jefferson Avenue
Newport News, VA 23606

Richard Hutson
MS H838, MP-5
Los Alamos National Laboratory
P.O. Box 1663
Los Alamos, NM 87545

Ken Jacobs
MIT Bates Linac
P.O. Box 846
Middleton, MA 01949

Gerald P. Jackson
MS 308
Fermilab
P.O. Box 500
Batavia, IL 60510

Marvin Johnson
MS 308
Fermilab
P.O. Box 500
Batavia, IL 60510

James B. Johnston
Lawrence Berkeley Laboratory
One Cyclotron Road
Berkeley, CA 94720

Kevin W. Jones
MP-6, MS H812
Los Alamos National Laboratory
P.O. Box 1663
Los Alamos, NM 87545

Ray Juras*
Oak Ridge National Laboratory
Oak Ridge, TN 37831

Kevin J. Kleman
Sychrotron Radiation Center
University of Wisconsin
3725 Schneider Drive
Stoughton, WI 53589

Peter Kloeppel
CEBAF
12000 Jefferson Avenue
Newport News, VA 23606

Shane R. Koscielniak
TRIUMF
4004 Wesbrook Mall
Vancouver, BC, Canada V6T 2A3

Stephen L. Kramer
Bldg. 360, C-237
Argonne National Laboratory
9700 S. Cass Avenue
Argonne, IL 60439

Bernard Kulke
L-627
Lawrence Livermore Laboratory
P.O. Box 808
Livermore, CA 94550

James R. Lackey
MS 341
Fermilab
P.O. Box 500
Batavia, IL 60510

Ronald F. Lankshear
Nuclear Energy Department,
Bldg. 902C
Brookhaven National Laboratory
Upton, NY 11973

Ronald Lauze
CEBAF
12000 Jefferson Avenue
Newport News, VA 23606

David M. Lee
MS H838, MP-5
Los Alamos National Laboratory
P.O. Box 1663
Los Alamos, NM 87545

Robert G. Malone
Nuclear Energy Department,
  Bldg. 902C
Brookhaven National Laboratory
Upton, NY 11973

Felix Marti
Cyclotron Laboratory
Michigan State University
East Lansing, MI 48824

Donald J. Martin
1004 Illinois Avenue
St. Charles, IL 60174

David McGinnis
MS 341
Fermilab
P.O. Box 500
Batavia, IL 60510

Jerry Melcher
Analytek
10261 Bubb Road
Cupertino, CA 95014

Roman Nawrocky
NSLS, Bldg. 725B
Brookhaven National Laboratory
Upton, NY 11973

Yat C. Ng
M/S A01-26
Grumman Corporation
Bethpage, NY 11714

Robert Odom
Charles Evans & Associates
301 Chesapeake Drive
Redwood City, CA 94063

Marvin Olson
MS 308
Fermilab
P.O. Box 500
Batavia, IL 60510

Paul V. Pancella
Indiana University Cyclotron Facility
2401 Milo B. Sampson Lane
Bloomington, Indiana 47408

Richard C. Pardo
Bldg. 203, Room F 153
Argonne National Laboratory
9700 S. Cass Avenue
Argonne, IL 60439

Ralph J. Pasquinelli
MS 341
Fermilab
P.O. Box 500
Batavia, IL 60510

Antanas V. Rauchas*
Bldg. 360
Argonne National Laboratory
9700 S. Cass Avenue
Argonne, IL 600439

David Pearce
TRIUMF
4004 Wesbrook Mall
Vancouver, BC, Canada V6T 2A3

William R. Rawnsley
TRIUMF
4004 Wesbrook Mall
Vancouver, BC, Canada V6T 2A3

John Perry
CEBAF
12000 Jefferson Avenue
Newport News, VA 23606

Joseph Rogers
NSLS, Bldg. 725B
Brookhaven National Laboratory
Upton, NY 11973

David Peterson
MS 341
Fermilab
P.O. Box 500
Batavia, IL 60510

Marc Ross*
Stanford Linear Accelerator Center
Stanford University
P.O. Box 4349
Stanford, CA 94720

Michael Plum
MP-5, MS H838
Los Alamos National Laboratory
P.O. Box 1663
Los Alamos, NM 87545

Robert Rossmanith
CEBAF
12000 Jefferson Avenue
Newport, News, VA 23606

John Power
MS H808
Los Alamos National Laboratory
P.O. Box 1663
Los Alamos, NM 87545

Jeffrey L. Rothman
NSLS, Bldg. 725D
Brookhaven National Laboratory
Upton, NY 11973

John C. Rothmann, Jr.
Nuclear Energy Department,
  Bldg. 902C
Brookhaven National Laboratory
Upton, NY 11973

Donald Priester
ER-90, GTN
US Department of Energy
Washington, DC 20545

Timothy Radway
Jorway Corporation
27 Bond Street
Westbury, NY 11590

Oscar Sander
MS H818
Los Alamos National Laboratory
Los Alamos, NM 87545

Robert Shafer*
MS H808
Los Alamos National Laboratory
P.O. Box 1663
Los Alamos, NM 87545

Thomas Shea
ADD, Bldg. 1005
Brookhaven National Laboratory
Upton, NY 11973

Brad Sherrill
Cyclotron Laboratory
Michigan State University
East Lansing, MI 48824

Stefan Simrock
CEBAF
12000 Jefferson Avenue
Newport News, VA 23606

William P. Sims
AGS Department, Bldg. 911B
Brookhaven National Laboratory
Upton, NY 11973

Gary A. Smith
AGS Department, Bldg. 911B
Brookhaven National Laboratory
Upton, NY 11973

Lief Solensten
M/S A01-26
Grumman Corporation
Bethpage, NY 11714-3580

Arnold N. Stillman
AGS Department, Bldg. 911B
Brookhaven National Laboratory
Upton, NY 11973

Gregory D. Stover*
Lawrence Berkeley Laboratory
One Cyclotron Road
Berkeley, CA 94720

Richard Talman
Wilson Laboratory
Cornell University
Ithaca, NY 14853

Gianni Tassotto
MS 222
Fermilab
P.O. Box 500
Batavia, IL 60510

Masami Torikoshi
Mitsubishi Electric Corp.
Wadasaki-cho 1-1-2
Hyogo, Kobe, Japan 652

Nicholas Tsoupas
Nuclear Energy Department,
    Bldg. 902C
Brookhaven National Laboratory
Upton, NY 11973

James Ulenas
Vetra Systems Corporation
1670 Old Country Road
Plainview, NY 11803

K. B. Unser
CERN, LEP Division
CH-1211 Geneva 23
SWITZERLAND

Olin B. van Dyck*
MP-DO, MS H844
Los Alamos National Laboratory
P.O. Box 1663
Los Alamos, NM 87545

Peter E. Vanier
Nuclear Energy Department,
    Bldg. 902C
Brookhaven National Laboratory
Upton, NY 11073

Gregory Vogel
MS 308
Fermilab
P.O. Box 500
Batavia, IL 60510

Gerald J. Volk
Bldg. 360
Argonne National Laboratory
9700 S. Cass Avenue
Argonne, IL 60439

U. von Wimmersperg
Nuclear Energy Department,
    Bldg. 902C
Brookhaven National Laboratory
Upton, NY 11973

Dan Y. Wang
Brobeck Division of Maxwell
    Labs., Inc.
1235 Tenth Street
Berkeley, CA 94710-1593

Tony Warwick
AFRD Mailstop 46-161
Lawrence Berkeley Laboratory
One Cyclotron Road
Berkeley, CA 34720

Robert C. Webber*
MS 308
Fermilab
P.O. Box 500
Batavia, Illinois 60510

Frank D. Wells
MS H808
Los Alamos National Laboratory
P.O. Box 1663
Los Alamos, NM 87545

Cliff Wiener
AGS Department, Bldg. 911B
Brookhaven National Laboratory
Upton, NY 11973

Carol Wilkinson
MS H838, MP-5
Los Alamos National Laboratory
P.O. Box 1663
Los Alamos, NM 87545

Stephen H. Williams
Bin 60
SLAC
2575 Sand Hill Road
Menlo Park, CA 94025

Richard Witkover*
AGS Department, Bldg. 911B
Brookhaven National Laboratory
Upton, NY 11973

Willie L. Wright, Jr.
US Army Strategic Defense Command
Huntsville, Alabama 35807

Emil P. Zitvogel
AGS Department, Bldg. 911B
Brookhaven National Laboratory
Upton, NY 11973

Robert A. Zolecki
Bldg. 360
Argonne National Laboratory
9700 S. Cass Avenue
Argonne, IL 60439

Martin Zucker
Nuclear Energy Department,
    Bldg. 902C
Brookhaven National Laboratory
Upton, NY 11973

*Organizing Committee Member

ACCELERATOR INSTRUMENTATION WORKSHOP

Brookhaven National Laboratory, Upton, New York
Berkner Hall–Building 488
October 23–26, 1989

Schedule

Sunday, 22 October

3:00 – 7:00 p.m.   Berkner Hall–*Registration and Sandwiches*

Monday, 23 October

9:15 – 9:30 a.m.	Room B–*Welcome: Derek Lowenstein, AGS*
9:30 – 12:30 p.m.	Room B–*Beam Intensity: Richard Talman, Cornell, SSC*
12:30 – 2:00 p.m.	Cafeteria–*Lunch*
2:00 – 5:00 p.m.	Room B–*Beam Position, Robert Shafer, LANL*
5:00 – 6:00 p.m.	Cafeteria–*Dinner*
6:00 – 7:30 p.m.	Lobby–*Cocktails*
7:30 – 9:00 p.m.	Room B: *Round Table Discussion:*
	How to promote inter-laboratory cooperation and industrial interest

Tuesday, 24 October

9:00 – 12:00 p.m.	Room B–*Beam Profile Measurement, John Galayda, NSLS*
Noon – 1:30 p.m.	Cafeteria–*Lunch*
1:30 – 4:30 p.m.	Room B–*Longitudinal Emittance, Robert Webber, FNAL*
5:00 p.m.	Lobby-Busses depart for *Tour of Lenz Winery* and *Lobster Dinner at Claudio's Restaurant*, Greenport

Wednesday, 25 October

9:00 – Noon	Theater: *Transverse Emittance, Oscar Sander, LANL*
Noon – 2:00 p.m.	Cafeteria–*Lunch*
2:00 – 5:00 p.m.	Theater–*Beam Loss Measurement, Marvin Johnson, FNAL*
5:00 – 5:30 p.m.	Theater–*Closing Remarks, Organization of next Workshop*

Thursday, 26 October

9:00 – Noon   *Tours of BNL Accelerator Facilities*

AUTHOR INDEX

AIP Conference Proceedings

		L.C. Number	ISBN
No. 119	Laser Techniques in the Extreme Ultraviolet (OSA, Boulder, Colorado, 1984)	84-72128	0-88318-318-8
No. 120	Optical Effects in Amorphous Semiconductors (Snowbird, Utah, 1984)	84-72419	0-88318-319-6
No. 121	High Energy e^+e^- Interactions (Vanderbilt, 1984)	84-72632	0-88318-320-X
No. 122	The Physics of VLSI (Xerox, Palo Alto, 1984)	84-72729	0-88318-321-8
No. 123	Intersections Between Particle and Nuclear Physics (Steamboat Springs, 1984)	84-72790	0-88318-322-6
No. 124	Neutron-Nucleus Collisions – A Probe of Nuclear Structure (Burr Oak State Park - 1984)	84-73216	0-88318-323-4
No. 125	Capture Gamma-Ray Spectroscopy and Related Topics – 1984 (Internat. Symposium, Knoxville)	84-73303	0-88318-324-2
No. 126	Solar Neutrinos and Neutrino Astronomy (Homestake, 1984)	84-63143	0-88318-325-0
No. 127	Physics of High Energy Particle Accelerators (BNL/SUNY Summer School, 1983)	85-70057	0-88318-326-9
No. 128	Nuclear Physics with Stored, Cooled Beams (McCormick's Creek State Park, Indiana, 1984)	85-71167	0-88318-327-7
No. 129	Radiofrequency Plasma Heating (Sixth Topical Conference, Callaway Gardens, GA, 1985)	85-48027	0-88318-328-5
No. 130	Laser Acceleration of Particles (Malibu, California, 1985)	85-48028	0-88318-329-3
No. 131	Workshop on Polarized ^{3}He Beams and Targets (Princeton, New Jersey, 1984)	85-48026	0-88318-330-7
No. 132	Hadron Spectroscopy–1985 (International Conference, Univ. of Maryland)	85-72537	0-88318-331-5
No. 133	Hadronic Probes and Nuclear Interactions (Arizona State University, 1985)	85-72638	0-88318-332-3
No. 134	The State of High Energy Physics (BNL/SUNY Summer School, 1983)	85-73170	0-88318-333-1
No. 135	Energy Sources: Conservation and Renewables (APS, Washington, DC, 1985)	85-73019	0-88318-334-X
No. 136	Atomic Theory Workshop on Relativistic and QED Effects in Heavy Atoms	85-73790	0-88318-335-8
No. 137	Polymer-Flow Interaction (La Jolla Institute, 1985)	85-73915	0-88318-336-6